LIFE IN STONE

FOSSILS OF THE COLORADO PLATEAU

by
Christa Sadler

GRAND CANYON ASSOCIATION
Grand Canyon, Arizona

DEDICATION

*To my mother and father, who gave me both a
love of the land and the ability to be out on it.
And to the Earth, whose stories come from her
heart. May we continue to listen.*

GRAND
CANYON
ASSOCIATION

Grand Canyon Association
P.O. Box 399
Grand Canyon, AZ 86023-0399
(800) 858-2808
www.grandcanyon.org

Reprint History 10 9 8 7 6 5 4 3 2 1

ISBN 0-938216-81-3
Library of Congress Control Number: 2004023554

Library of Congress Cataloging-in-Publication Data

Sadler, Christa.
    Life in stone : fossils of the Colorado Plateau / Christa Sadler.--
1st ed.
       p. cm.
    Includes bibliographical references and index.
    ISBN 0-938216-81-3
  1. Fossils--Colorado Plateau. 2. Geology--Colorado Plateau. 3.
Geology, Stratigraphic. I. Title.
    QE79.5.S23 2005
    560'.9788'1--dc22
                                                                 2004023554

Edited by Sandra Scott
Designed by Christina Watkins and Amanda Summers
Indexed by Earl E. Spamer
Printed in China by Global Interprint, Inc. on recycled paper using
vegetable-based inks.

PHOTOGRAPHY CREDITS

Lee Udall Bennion 68; Bureau of Land Management, Salt Lake City Office 28
right; Bureau of Land Management, Grand Staircase–Escalante Office 44 middle,
51 middle (both); Michael Collier 24b, 39, 52, 57t, 64; Denver Museum of
Nature and Science cover background, 14, 45t, 46t, 54–55, 55t, 56, 57b; David
Edwards 21t middle; William E. Ferguson 13t left; Grand Canyon National Park
Museum Collection 3 fern, 9b, 10 #8323, 19 #10936, 20 #4420B, 21t right
#11096B, 27 right #21372C (preceding by Michael Quinn for NPS), 58 #7174,
#0793, #9374A, 61b #6992 and 6990 (both by Michael Bobko for NPS);
George H. H. Huey cover inset, 8, 23t, 26t left, 29, 32t, 32 middle, 44–45, 45,
53, 60; Gary Ladd 63; Jim Mead 59; Steve Mulligan 30; Museum of Northern
Arizona 15, 21 middle, 22t, 27 left, 32b left, 33b, 34b, 44t, 55b; Christa Sadler
9t and middle, 11, 16, 17, 21t left, 23b, 24t right, 25, 26t right, 31, 34t, 50,
51t; William D. Tidwell 21b, 41, 43, 46b, 49, 70; Stephen Trimble 26b left, 28
middle, 36, 40 (both), 47; Utah Geological Survey 28 left; Utah Museum of
Natural History 48, 51b left; Shuhai Xiao 13 right, Bob Young 24t left.

ILLUSTRATION CREDITS

Ron Blakey 12m, 16, 20t left, 22b, 27t, 32, 37, 42b, 53; Trustees of the British
Museum (Natural History) 26, 49, 56; Chalk Butte, Inc. 11; Denver Museum of
Nature and Science 12t, 19, 32, 33t, 36; David Elliott 18; Intermountain Natural
History Association 36t; Museum of Northern Arizona (MNA) 15 middle, 23
and 35 (both by Pamela Scott Lungé for MNA), 27b, 34–35, 60; Deborah Reade
inside front cover (icons), 4; Paleomap Project, C. R. Scotese 12-13b, 31;
Amanda Summers inside front cover, 7, 72, inside back cover; William D.
Tidwell 42t, 43; Joe Tucciarone 52; Carol Brist Von Kemper 39; Wood Ronsaville
Harlin, Inc. (for Grand Canyon Association) cover, 22t, 25t.

*Grand Canyon Association is a nonprofit educational organization. Net
proceeds from the sale of this book will be used to support the education
and research goals of Grand Canyon National Park.*

TABLE OF CONTENTS

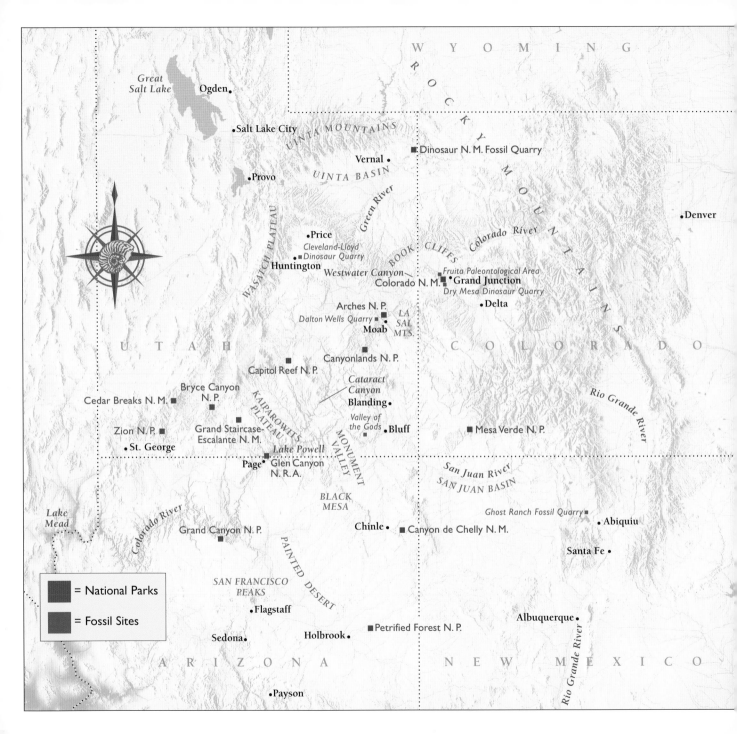

WYOMING

*Great
Salt Lake*

•Ogden

•Salt Lake City

UINTA MOUNTAINS

■ Dinosaur N. M. Fossil Quarry

Vernal •

UINTA BASIN

•Provo

ROCKY MOUNTAINS

•Denver

•Price

*Cleveland-Lloyd
Dinosaur Quarry* ■

Huntington •

Green River

BOOK CLIFFS

Colorado River

Westwater Canyon —

Colorado N. M. ■

Fruita Paleontological Area ■
■ •Grand Junction
Dry Mesa Dinosaur Quarry

Arches N. P. ■

Dalton Wells Quarry ■

*LA
SAL
MTS.*

•Delta

Moab •

U T A H

C O L O R A D O

Canyonlands N. P. ■

Capitol Reef N. P. ■

*Cataract
Canyon*

Blanding •

Rio Grande River

Cedar Breaks N. M. ■

Bryce Canyon
N. P. ■

*KAIPAROWITS
PLATEAU*

*Valley of
the Gods* ■

•Bluff

■ Mesa Verde N. P.

Zion N. P. ■ ■

Grand Staircase-
Escalante N. M. ■

*MONUMENT
VALLEY*

•St. George

Lake Powell

Page • Glen Canyon
N. R. A.

San Juan River
SAN JUAN BASIN

*BLACK
MESA*

*Lake
Mead*

Colorado River

Grand Canyon N. P. ■

PAINTED DESERT

Chinle •

■ Canyon de Chelly N. M.

Ghost Ranch Fossil Quarry ■

•Abiquiu

Santa Fe •

*SAN FRANCISCO
PEAKS*

•Flagstaff

Albuquerque •

Sedona •

Holbrook •

■ Petrified Forest N. P.

Rio Grande River

A R I Z O N A

N E W   M E X I C O

•Payson

■ = National Parks

■ = Fossil Sites

The map at left shows the locations of national parks, monuments, and fossil quarries that are mentioned in this book. Towns and selected geographic features are shown for reference, but roads have been omitted.

THE COLORADO PLATEAU

The Colorado Plateau is an extraordinary region. A vast and colorful landscape of eroded canyons, impressive buttes, whimsical hoodoos, and grand cliffs, this area of the southwestern United States encompasses 130,000 square miles of land, larger than New York, Ohio, and Pennsylvania combined. At an average elevation of five thousand feet, this semi-arid land exposes a greater number and diversity of erosional landforms than anywhere else on the planet. Its rock layers span almost two billion years of earth history. Travel on the Colorado Plateau is travel through time; millions of years of earth history are laid bare for examination.

The layers of sedimentary rock that are exposed on the plateau like the pages of a book represent a diversity of environments and hundreds of millions of years. Contained within these rocks are the remains of the organisms that lived in these environments. The stories of swamps and oceans, great slow-moving rivers and wind-blown sand dunes, tidal flats and tropical seas come to life as we unlock the secrets preserved in these pages.

Paleontological research began on the plateau in the late 1800s, the purview of a few flamboyant personalities. While techniques have improved and tempers have (sometimes) cooled, research continues. In fact, there has been a renaissance in Colorado Plateau paleontology in recent years. The focus has shifted from the "trophy hunting" of years past to greater analysis and assessment of such questions as biodiversity through time, evolutionary relationships, climate shift, and environmental change. The region is recognized as one of the finest earth science laboratories in the world, and the new discoveries and analyses made here are answering questions, solving mysteries, and making connections that help us understand the history of life worldwide.

You've been hiking across the desert in northeastern Utah, through rounded badland hills, and you see something intriguing protruding from the grey shale slope. Stepping closer, you discover the corner of an old bone. It is hard and dark and looks like nothing you've seen before.

Walking the rim of the Grand Canyon, you realize that you are standing on shells and corals encased in cream-colored rock.

In a remote wash on the Navajo Reservation in northern Arizona, you find giant logs embedded in soft shale, so perfectly preserved that the patterns of bark, tree rings, and knots remain.

In each of these cases you have found the preserved remnants of ancient life: fossils. Your next thought might be, "I may be the first person to see this since it died millions of years ago!" The thrill of your initial discovery will continue as your find is properly excavated and analyzed. You learn that the bone was from a turtle that lived 50 million years ago on the shore of a huge lake. The shells and corals lived in a clear, warm, tropical sea 270 million years ago, and the wood came from 100-foot-tall conifers that were swept down a flooding river over 200 million years ago.

Fossils are any preserved remains or evidence of ancient life, and new ones are being discovered every day. Plants, animals, even bacteria and fungi can fossilize. Fossils can be bones, wood, shells, leaves, impressions, tracks, and feces. All of these remains can tell us how organisms moved, ate, reproduced, lived, died, and ultimately evolved. We can learn how different groups of organisms were related to one another and what a region was like millions of years ago.

A fundamental idea in earth science is that the processes and laws that govern today's natural world operated in the past as well, and by observing the modern world we can unearth clues about the past. So if we look at certain life forms in modern environments, we can get a good idea of how ancient organisms once lived.

Four centuries ago most people believed that the earth was no more than six thousand years old, a figure obtained by calculating the generations in the Book of Genesis. As knowledge of the world grew, people finally accepted that fossils are the remains of once-living organisms. In the mid-nineteenth century the development of evolutionary theory finally paved the way for the acceptance of fossils as organisms that had lived and died as part of a natural cycle of organic evolution and extinction.

The study of fossils has much to teach us about the world in which we live. The development of life from simple, single-celled bacteria to the diversity of creatures we see today is a marvelous story full of complexities, questions, and mysteries. The study of past life enlightens us about our place in the web of life, as well as that of other life forms with which we share this planet. Paleontology gives us insights into which evolutionary adaptations in earth history have succeeded, those that have failed, and how far we may be able to push our present ecosystems before they, too, respond with failure.

MAKING A FOSSIL

Not everything that dies becomes a fossil. Conditions must be just right. Hard substances such as bone, teeth, shell, chitin (a material similar to your fingernails), or wood are far more likely to become fossils than soft-bodied organisms such as worms, jellyfish, or butterflies.

In order to become fossilized, a plant or animal must be covered by sediment or otherwise protected from scavenging and decay almost immediately after it dies (A,B). Once buried, the organism may undergo the same changes that turn soft sediment into rock (C): minerals work upon the hard parts of

when silica-rich ground water permeates buried wood and silica molecules fill in pore spaces. This happens so slowly that fine details of bark and tree rings are preserved. Permineralized wood and bones are heavy because they have literally "turned to stone." Sometimes another mineral, such as pyrite (fool's gold) or jasper (red chert) will replace the minerals in an already fossilized object.

Plants and soft-bodied animals may be preserved when, under the pressure of burial, all the liquids and gases that form a living organism are forced away, leaving only a two-dimensional image in carbon, the basis of all life.

Quite often the original organism is gone, but its shape and detail are preserved. If the organism erodes out of hard sediment, the hole it leaves is called a mold. If this hole becomes filled with sediment, the result is a cast. Despite the lack of original material, casts and molds are important in determining structure and shape of ancient creatures.

Sometimes only a trace of the original organism is preserved. Tracks, burrows, feeding traces, and coprolites (fossilized feces) provide clues to the behavior and diet of extinct organisms that can't be obtained from body fossils alone.

mammoths and pack rat nests that are merely thousands of years old.

Chances are slim that an object will be preserved as a fossil; those found so far represent only a tiny portion of past life. Biologists estimate that there are roughly 4,500,000 species living on Earth today, and in all our years of searching, we've found fewer than three hundred thousand fossil species from 3.5 billion years of life history!

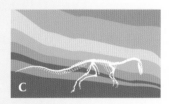

the organism in several different ways. In a process called permineralization, minerals can completely replace the original material of the organism or fill in any empty spaces. Petrified wood forms

Insects encased in amber (fossilized tree sap) for millions of years look as fresh as the day they died. On rare occasions freezing and desiccation can preserve soft tissue.

How old must something be to be called a fossil? Some paleontologists feel that something must be at least a million years old to be called a fossil. Other scientists work with "fossils" of frozen

THE SURROUNDING ROCKS

The rocks of the Colorado Plateau may appear too numerous and diverse to make sense of, with all of their colors, textures, shapes, and consistencies. In reality all rocks can be categorized into three basic classes—igneous, metamorphic, or sedimentary—based on how they formed. Within each class there are several types, and these too can be divided into understandable groups.

Sedimentary rocks are the most important for paleontologists; except under very unusual circumstances these are the only rocks in which fossils are found. Sedimentary rocks are fundamentally easy for us to understand, because they form in familiar settings like lakes, rivers, oceans and deserts, and so they provide a snapshot of how Earth's surface environments looked in the past.

Some sedimentary rocks form through the breakdown of other rocks into particles of various sizes: cobbles, gravel, sand, silt, and clay. Water, wind, or ice will move these fragments from their source area, often carrying them hundreds of miles. When particles can no longer be transported they are deposited as soft sediment. These sediments accumulate layer upon layer over hundreds of thousands and millions of years, and the weight of overlying sediment compacts the layers beneath. Minerals (calcium carbonate, silicon dioxide, and iron oxide) fill the air spaces between the grains and cement them together.

Sedimentary rocks can also form from plant or animal remains. Coal is composed of compacted plant material. In the ocean, calcium carbonate or silica from plants and animal shells accumulate in layers to be compacted into sedimentary rocks such as chalk, chert, or limestone. Certain chemicals—gypsum, salt, and some forms of chert and limestone—can precipitate directly from water to form rock.

The sedimentary layers (called formations) on the Colorado Plateau were deposited in environments such as lakes, riverbeds, beaches, shallow seas, deserts, floodplains, and tidal flats. Each of these environments has left characteristic features in the sediment that tell of the conditions under which the rock formed. Ripple marks indicate water currents. Mud cracks tell of alternating wet and dry conditions. Red color in some layers means there was oxygen present to oxidize the iron in the sediment. The white and green of other formations indicate oxygen-poor conditions at times. Grey is evidence of organic matter.

Living creatures inhabited all these environments, and many have become the fossils that tell us stories of the plateau's past.

Because it may take many hundreds of thousands or millions of years to deposit one layer of sedimentary rock, it is important to remember that every formation may represent a long time period during which environmental conditions and living organisms may change greatly. Therefore, any discussion of ancient environments and creatures of a particular formation will by necessity sound more static than real life.

Fossil mudcracks in shale

Ripples preserved in sandstone

Fossil sponge in limestone

DEEP TIME

THE PRECAMBRIAN

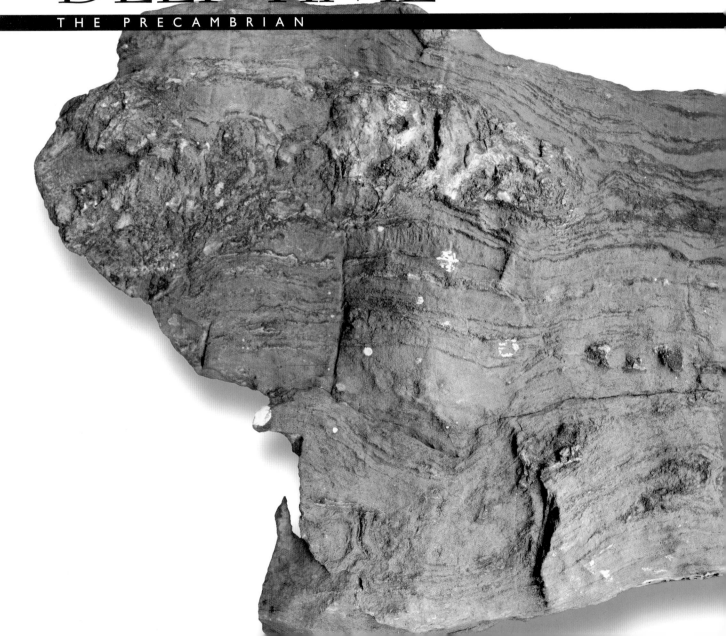

The Precambrian, the time during which the primeval continents, oceans, and atmosphere were forming, encompasses 87 percent of the earth's history. Around 3.5 billion years ago, the planet's first lifeforms appeared. Microscopic single cells without nuclei began to flourish, possibly around volcanic hot springs that dotted the oceans. In shallow intertidal waters, these cyanobacteria occasionally formed wide-spread mats in which lime precipitated and sediment collected. The cyanobacteria then grew through the mat to the surface, trapping more sediment, and so on. The result was a multilayered cyanobacteria and lime "club sandwich" called a stromatolite.

Cyanobacteria are more important than casual observation might suggest. These creatures were the first life able to produce oxygen through photosynthesis. Until then, Earth's atmosphere lacked free oxygen, containing instead greater proportions of carbon dioxide, nitrogen, water vapor, sulfur gases, and possibly methane. Over time, the atmosphere slowly became enriched with oxygen, but it took a long time to reach levels that would have permitted complex life to form. As recently as 600 million years ago, there still may have been only 10 percent of modern levels of oxygen in the atmosphere.

The record of the Precambrian on the Colorado Plateau begins about 1.8 billion years ago. Rocks from this period are exposed at the bottom of Grand Canyon, in the Colorado River's Westwater Canyon, and on the eastern edge of the plateau in Colorado's Black Canyon of the

Several of the later Precambrian formations of the Grand Canyon Supergroup preserve fossilized algal stromatolites, "reefs" several feet thick and dozens of feet long. Their undulating layer-cake construction is clearly visible (opposite page).

Gunnison River. These are highly deformed metamorphic rocks that resulted from continuous collisions between crustal plates during the formation of the North American continent. They preserve no fossil remains, although stromatolites surely existed in the shallow waters of this ancient ocean.

By 1.2 billion years ago, the plate collisions in the plateau region had long since quieted and erosion had planed the terrain to near sea level. The region that would become the Colorado Plateau was covered with sediments deposited in shoreline and shallow marine environments. These sediments that covered the eroded schist are called the Grand Canyon Supergroup, the only sediments of this age exposed on the plateau.

Life was still simple, but some innovations had evolved. Single-celled plants and animals had appeared and had become quite successful, diversifying beyond their primitive ancestors. Although simple, these fossils can be extremely numerous in the sediments of the Grand Canyon Supergroup. A cubic centimeter of the shales in one particular layer can contain ten thousand algal or bacterial cysts! Trace

In addition to single-celled creatures, the later Precambrian seas of the Colorado Plateau also may have held creatures similar to those shown in this scene from Precambrian Australia (above).

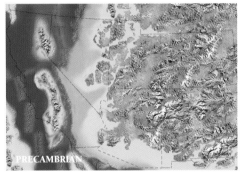

PRECAMBRIAN

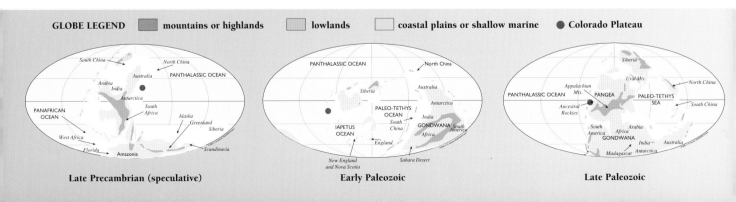

GLOBE LEGEND ▢ mountains or highlands ▢ lowlands ▢ coastal plains or shallow marine ● Colorado Plateau

Late Precambrian (speculative)

South China — North China
Australia
Arabia PANTHALASSIC OCEAN
India
Antarctica
PANAFRICAN
OCEAN South
Africa Alaska
Greenland
West Africa Siberia
Florida Amazonia Scandinavia

Early Paleozoic

PANTHALASSIC OCEAN North China
Australia
Siberia Antarctica
PALEO-TETHYS
OCEAN
South
China India GONDWANA South
America
IAPETUS Africa
OCEAN
England
New England Sahara Desert
and Nova Scotia

Late Paleozoic

Siberia
Appalachian Ural Mts.
Mts. North China
PANTHALASSIC OCEAN PANGEA
Ancestral PALEO-TETHYS South China
Rockies SEA
South Arabia
America Africa India Australia
GONDWANA
Madagascar Antarctica

Shark Bay in western Australia is a setting similar to that of the later Precambrian period on the Colorado Plateau, where abundant stromatolites grow in the calm, shallow, intertidal waters of a warm, enclosed bay.

3x actual size

Chuaria circularis from Grand Canyon's later Precambrian sediments has kept paleontologists guessing for more than a century. Sometimes interpreted as an algal cyst, it may represent a life-form that has no modern descendants.

fossils interpreted as impressions of stranded jellyfish or burrows of wormlike organisms have also been found in the Grand Canyon Supergroup. Multicellular animals may have made these traces. If this is true, they remain among the earliest such fossils known.

These primitive fossils do tell us about the environment of the Colorado Plateau area during late Precambrian times. The stromatolites and other algal remains suggest quiet, shallow, and possibly extremely saline waters.

The world of the Colorado Plateau region one billion years ago was a barren expanse of rock. Probably no life existed on land. There were no green plants, no insects, no fish in the seas. The ocean teemed with microscopic single-celled plants and animals. While many of these cells would remain in their primitive state, some would soon take an important step in the development of life on earth.

MOBILE EARTH

Our planet has not always looked the way it does today. The shape and location of continents have been changing continually since they first began to form at least four billion years ago. Earth's lithosphere (crust) is broken into several pieces like a great jigsaw puzzle. These pieces, called plates, move about in relation to one another, alternately colliding, separating, or sliding past each other. The continents ride along on these plates like boxes on a conveyor belt, and wherever the plate goes, so goes its continental cargo. Several times in Earth's history the continents have joined into one gigantic supercontinent, the latest of which is called Pangea ("all world"). Pangea formed about 245 million years ago and began to split apart not long after. By 20 million years ago the continents were arranged much like they are today, although they continue to move.

Throughout its history, the area that would become the Colorado Plateau has changed position in many ways. Changes over time in latitude, longitude, and elevation above or below sea level are all reflected in the rocks and fossils preserved here.

Mesozoic

Early Cenozoic

Late Cenozoic

BRANCHING OUT

Cambrian trilobite
from central Utah

The Cambrian seas of Grand Canyon were home to several kinds of trilobite, whose closest living relative is the modern horseshoe crab.

During the Paleozoic, the area that would become the Colorado Plateau was a huge coastal plain, across which oceans advanced and retreated over time. Mud and sand deposited from eroding mountain ranges to the east periodically washed over this plain, especially later in the era.

At the beginning of the Paleozoic Era, evolution took an astonishing turn. From single-celled creatures and few more complex organisms, life expanded into an array of forms and adaptations. The Paleozoic Era began with a tremendous diversification of life forms and ended with the largest mass extinction in Earth's history.

THE CAMBRIAN EXPLOSION

During the Early Cambrian Period, multicellular life proliferated at an astonishing rate, geologically speaking. Members of almost every major animal group appeared in the oceans in a relatively short period, perhaps as little as five million years. Early members of major invertebrate groups such as brachiopods, mollusks, sponges, corals, echinoderms, and arthropods emerged during this time period. Even primitive vertebrates appeared; instead of a true backbone, they had only a muscular nerve canal or "notochord" running down their backs.

What accounts for this rapid expansion of life forms and why at this time? Ideas range from increasing oxygen levels in the atmosphere to changes in ocean chemistry and circulation. Perhaps these creatures had been quietly evolving in corners and crevices of the Precambrian world, and their "sudden" appearance is really only because the fossil record hasn't preserved earlier

Trilobites left their tracks in the mud of the Bright Angel Shale over 500 million years ago.

¼ actual size

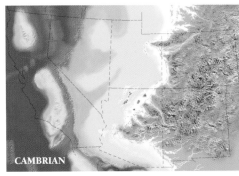

CAMBRIAN

Grand Canyon's Tapeats Sandstone (above), and Muav Limestone (at bottom) record the advance of an ancient sea during the Cambrian Period.

remains. Research in Grand Canyon's Cambrian-aged layers suggests the possibility that a wide diversity of environmental conditions in Cambrian seas may have allowed these early animals to evolve more quickly as they adapted to broad variations in such factors as topography, ocean circulation, salinity, temperature, and ocean depth.

What we do know is that by around 530 million years ago the sea floor would not have seemed as barren of life as it had a few million years earlier. The great majority of the inhabitants were trilobites, whose closest living relatives are modern horseshoe crabs. Trilobites had segmented bodies, jointed legs, and exoskeletons made of chitin. Their bodies were divided into three lobes (hence their name), and some could roll themselves up like pill bugs to avoid predators. Since they molted as they grew, the great abundance of trilobite fossils is probably due in large part to the fact that we are finding their shed exoskeletons as fossils. All trilobites lived in marine environments but had a variety of lifestyles, from burrowing in mud to paddling and floating in water.

Fossils from the Cambrian formations of the Colorado Plateau illustrate the results of this remarkable diversification. The sediments indicate a shallow sea advancing from west to east over the continent. Cambrian-aged sediments are exposed in Grand Canyon as the Tapeats Sandstone, Bright Angel Shale, and Muav Limestone. These sediments record a beach-and-coastal environment, an offshore plain, and carbonate banks where sponges and brachiopods lived in the limey sediment.

In these 505- to 515-million-year-old sediments, trace fossils are the most common fossils. In places, the sandstone and shale are covered with thousands of burrows made by creatures feeding in and above the sediment. Vertical tubes may have been dwellings or resting places for organisms that were filtering food from the water.

More than forty species of trilobites have been found in the Bright Angel Shale. Usually, we find their moving and feeding traces left in the soft sediments of the sea floor. Mysterious fossils have also been recovered from these sediments; there are body and trace fossils that show connections to some of the enigmatic creatures discovered in the spectacular Burgess Shale of British Columbia. In fact, five hundred million years ago the two areas lay quite close to one another at the same latitude, and so may have shared some of the same creatures.

We also find brachiopods, mollusks, snails, sponges, and algae in these Cambrian sediments. Researchers have discovered compelling evidence of land-plant spores, perhaps washed into the ocean from the nearby continent. If these are indeed spores from land plants, it is the earliest evidence of terrestrial plants anywhere on the planet.

Life in the Cambrian consisted mostly of creatures that fed directly from the sediment, and these "deposit feeders" were very successful in the early development of marine communities. But as circulation in the oceans increased around the continents, organisms that filtered food out of the water currents began to outcompete the deposit feeders. While deposit feeders continued to exist, filter feeders came to characterize the oceans of the later Paleozoic.

Wormlike creatures bulldozing through the mud and ingesting microscopic food particles probably made the numerous finger- or cigar-shaped burrows found in the Grand Canyon's Bright Angel Shale and Tapeats Sandstone.

THE MISSING YEARS—THE ORDOVICIAN AND SILURIAN

During Ordovician and Silurian times, fish diversified in the world's oceans. The earliest fish had no jaws, and fed by sucking water and food through mouth slits in their dense, bony armor. Jaws developed over time, either from a set of arches supporting the gills or from another structure which functions in respiration and feeding in some more primitive forms.

Although excellent records of this time exist to the west

Fossil fish found in the Colorado Plateau's Devonian sediments lived a diversity of lifestyles in fresh, brackish, and marine waters. Some had rows of crushing teeth for eating shellfish. Others had no teeth at all, but probably strained plankton out of the water. Still others scavenged leftovers in the sediment and water or, like *Eldenosteus* (center right), were predators.

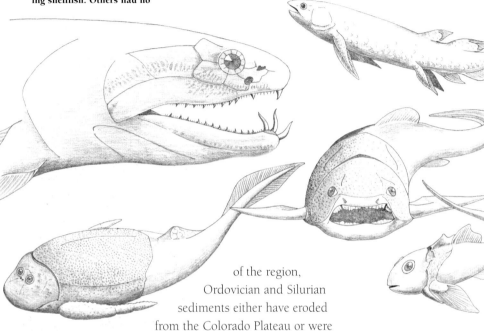

first animals with backbones thrived and diversified. Late Devonian sediments exist in Grand Canyon, near Flagstaff, Arizona, and at the southern edge of the plateau. These sediments, although not widespread, have yielded important clues to the nature of Devonian life on the Colorado Plateau.

In a tiny pocket of sediment no more than thirty feet square, uplifted against the eastern flank of Mount Elden near Flagsaff, about ten different species of ancient fish have been unearthed, including one form known worldwide only from this location. *Eldenosteus arizonensis* is a fish from an ancient group called the placoderms, meaning "plated skin." Only six-to-eight inches long, little *Eldenosteus* had two armor plates over its head and the front portion of its body, attached with a special ball-and-socket joint that allowed for movement and gill respiration. A serrated bone in its jaw served in place of teeth, and shows that it was a predator.

The plateau's Devonian fish ranged from a few inches to three or four feet in length. One fossil fish from this area is similar to a form that has been found worldwide, from Antarctica to the Arctic and Greenland. The presence of corals, sponges, brachiopods, and gastropods among these sediments indicates an ocean filled with diverse life.

of the region, Ordovician and Silurian sediments either have eroded from the Colorado Plateau or were never deposited, leaving us with no insights into the evolution of local marine communities during this time.

THE AGE OF FISHES—THE DEVONIAN

During the Devonian Period, fish and sharks proliferated in the world's oceans and nearshore environments. The region that would become the Colorado Plateau (which lay near the edge of a shallow sea during the latter half of the Devonian) was no exception. The shoreline retreated and advanced across the region, leaving alternately marine-to-brackish estuarine sediments. In these environments, the

A SEA-COVERED CONTINENT—THE MISSISSIPPIAN

Marine communities became more complex and varied throughout the Paleozoic. During the Mississippian a clear and warm, shallow, tropical sea covered most of what would become the United States, from Nevada across Kentucky. The well-preserved fossils in the Redwall Limestone paint a detailed picture of typical marine life during the middle Paleozoic. Most abundant are brachiopods and cornucopia-shaped horn corals. These corals did not form true reefs, but instead lived individually on the sea floor and waited for small prey to swim near their stinging cells or joined other members of the marine community in reeflike structures called bioherms.

Also common were crinoids, which are related to modern sea urchins and sea stars. As their common name "sea lilies" suggests, they resembled plants more than animals. A long, segmented stalk held up a complex head with many anemone-like arms that filtered food from the water currents. When the animal died its stem sections, looking like tiny lifesavers, were preserved in the sediment. Individual pieces of the stem are among the most common fossils found in marine rocks of Mississippian age on the plateau.

Like crinoids, bryozoans also resembled plants. These organisms are commonly called "moss animals" due to their lacy appearance, and their fossils are some of the loveliest from the plateau's Mississippian sediments. The organisms lived in pinpoint holes among the latticework and filtered food out of the water circulating nearby.

4x actual size

Shallow seas of the middle Paleozoic Colorado Plateau held a variety of life-forms indicating that warm tropical conditions prevailed. At left are fossil crinoid stem sections.

Sediments formed in the Mississippian seas of the Colorado Plateau hold the remains of crinoids (right) and delicate bryozoans (below).

4x actual size

MISSISSIPPIAN

Predators in this ocean included sharks and spectacular, squidlike nautiloids. These relatives of the octopus were the first true swimmers and predators and appeared relatively early in the Paleozoic. One species preserved in Grand Canyon is a cone-shaped nautiloid called *Rayonnoceras*. Its shell was divided into several chambers and the animal lived in the largest, outermost chamber. By changing the amount of air in posterior chambers to adjust its buoyancy it could move up or down the water column. Some individuals of this group reached seven to eight feet in length.

Many of the organisms found in this sea were filter feeders, so the water must have been clear and well aerated. Too much sediment in the water would clog filters and lack of current would not bring food to those creatures attached in one place. The Mississippian Colorado Plateau must have looked like a smooth, endless Caribbean with odd, fanciful creatures drifting through the water, corals and crinoids waving from the sea floor, and sharks and long, deadly nautiloids shooting here and there in search of trilobites or fish.

LEAVING THE WATER

As early as the Ordovician Period, and possibly even as far back as the Cambrian, the first primitive land plants had invaded the moist near-shore regions of the continents. They were quickly followed by scorpions and spiders, then by centipedes and other arthropods. Some aquatic animals had already developed limbs with fingers and toes by the Devonian, and by the Late Devonian some fish had primitive

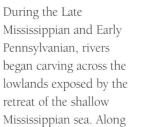

During the Late Mississippian and Early Pennsylvanian, rivers began carving across the lowlands exposed by the retreat of the shallow Mississippian sea. Along

The fossils of Grand Canyon's Redwall Limestone include (l to r) cone-shaped nautiloids related to modern squids, colonial corals, and solitary horn corals.

these waterways, sluggish swampy areas supported a wide variety of plants. Impressions of bark and roots from a primitive tree called *Lepidodendron*, the scale tree, were discovered in the Surprise Canyon Formation in Grand Canyon. The closest living relative of this giant of the Paleozoic swamps is the diminutive modern club moss. Other plant remains from the Pennsylvanian rocks of Grand Canyon include the early conifer *Walchia*, the seed ferns

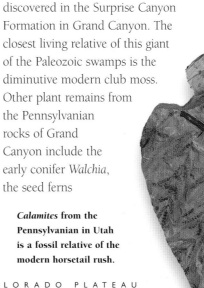

lungs and sturdy, stumpy fins that allowed them to spend extended amounts of time out of the water, shifting from pool to pool for spawning or seasonal movement, breathing through lungs instead of gills. The earliest true amphibians ventured forth early in the Mississippian, clumsy things with big heads and long tails. Life in the sea continued, but on land life was blossoming.

LIFE ON LAND—THE PENNSYLVANIAN

Tracks of terrestrial vertebrates and fossils of land plants in Pennsylvanian-aged sediments are among the first signs of the expansion of terrestrial life on the Colorado Plateau.

Calamites from the Pennsylvanian in Utah is a fossil relative of the modern horsetail rush.

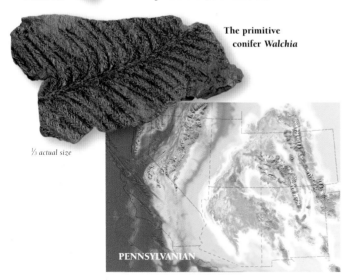

A scene from Pennsylvanian Utah

Impressions of the bark of
Lepidodendron, the "scale tree"

⅓ actual size

The primitive
conifer *Walchia*

⅓ actual size

PENNSYLVANIAN

Neuropteris and *Lygenopteris,* and primitive horsetail rushes. These plants grew on expansive floodplains and in small, ephemeral swamps near streams, and they indicate a warm and seasonally moist climate in this part of the plateau.

The Supai Group in Grand Canyon records a variety of environments from shallow seas to the river floodplains and wind blown dunes of a coastal plain. Across the floodplains and dunes wandered small reptiles and amphibians about the size of modern house cats. Their tracks suggest animals with heavy bodies and short legs.

Similar tracks as well as skeletal remains have been recovered from Pennsylvanian and Early Permian formations in southeastern Utah. Bones of amphibians have been found in formations exposed in the Monument Valley, Valley of the Gods, and San Juan River areas. These fossils suggest that the Pennsylvanian environment in this region was one of rivers, floodplains, and perhaps deltas. Of the amphibian remains found, by far the most common is a

form known as *Eryops,* a squat-bodied, flat-headed fellow up to ten feet in length. The remains of primitive lungfish and the more terrestrial amphibian *Seymouria* found in the sediments just south of Moab, Utah, suggest that there may have been local areas with extreme seasonal variation in water availability.

Elsewhere in the region, shallow seas still inundated the land. The Pennsylvanian Naco Formation at the plateau's southern edge near Payson, Arizona, contains one of the best shark faunas on the plateau. Researchers have found the same group of animals in Colorado, Iowa, and along the border of Wales and England, suggesting a continuous sea between all these places.

Limestones of the Paradox Formation in

Pennsylvanian- and Permian-aged sediments in Utah's Monument Valley record river floodplain and windblown desert environments.

¹⁄₁₀ actual size

The Late Paleozoic amphibian *Seymouria* (above)

The giant amphibian *Eryops*, from southeast Utah (left)

FOSSILS OF THE COLORADO PLATEAU

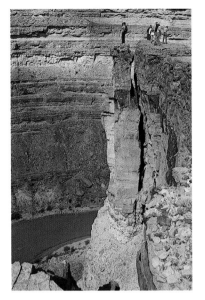

Clockwise, from left: The Goosenecks of the San Juan River; limestones of the Honaker Trail Formation; brachiopod in limestone replaced with jasper

southeastern Utah contain fossils that are valuable economically as well as scientifically. In Pennsylvanian times, the shallow edges of a long, narrow seaway that stretched across southwestern Colorado and southeastern Utah were perfect for accumulation of algae composed of calcium carbonate (lime). These tiny organisms loved a good warm spot with lots of light. Under these conditions, they grew layer upon layer, like miniature heads of lettuce. When they died, the "leaves" fell apart and accumulated as porous algal bioherms. Oil derived from nearby organically rich shales migrated into the air spaces in these fossilized "reefs," resulting in substantial deposits of oil in this region. One can see it seeping out of the limestone walls along the lower San Juan River in southeastern Utah.

The Honaker Trail Formation and other Late Pennsylvanian sediments also contain abundant marine fossils in the form of brachiopods, crinoids, corals, and calcareous algae. Some of the best exposures of these fossils are in the walls of the lower canyon of the San Juan River and in the Colorado River's Cataract Canyon in Canyonlands National Park. In many cases, these shells have been replaced during fossilization with jasper. The resulting blood-red fossils stand out clearly against the dull gray limestone.

END OF AN ERA—THE PERMIAN

At the dawning of the Permian Period, the earth had changed dramatically. Pangea was almost completely assembled and an ice cap covered the southern polar region of the supercontinent, making the global climate somewhat cooler, more seasonal, and increasingly arid. The seas had largely retreated from the Colorado Plateau, exposing a broad coastal plain. Across these lowlands the wind blew sand into dune fields, and huge sluggish rivers meandered, carrying billions of tons of fine sediments derived from the eroding mountains to the east. Plants with seeds were becoming dominant. Seeds require less moisture than spores, and so were better adapted to the arid to semi-arid climate.

Near Canyonlands National Park paleontologists have uncovered an ancient "logjam" of massive petrified conifer trunks, the largest trees known from the Permian Period. There are only six petrified logs known from this time in North America and three of them are on display at the Dinosaur Museum in Blanding, Utah. These logs also show marks that may be beetle and termite damage. If so, this would be the earliest evidence of such creatures known from the fossil record.

The broad floodplains, coastal dunes, tidal flats, and vast sand deserts of southeastern Utah, northeastern Arizona, and northwestern New Mexico were home to a variety of land-dwelling life forms in the Early Permian. Some of the more common reptiles include the fearsome predators *Ophiacodon*, *Sphenacodon*, *Dimetrodon*, and

The sail-finned predator *Dimetrodon* wandered the Permian forests and river floodplains now known as Valley of the Gods near southeastern Utah's Cedar Mesa.

Arthropod tracks in northern Arizona's Coconino Sandstone (bottom and above right) are almost identical to those found on modern sand dunes (above left).

the plant-eater *Edaphosaurus*.

All of these creatures belong to a group of animals that shared skeletal and dental characteristics of both mammals and reptiles (another member of the mammal-like reptiles would go on to be the progenitor of all mammals). On their backs, *Dimetrodon* and *Edaphosaurus* had a large fin of skin supported by lengthened vertebrae that stood up like a sail, which probably served to collect and dissipate body heat.

These fossils reveal much about terrestrial life during the Permian. Reptiles were now the dominant creatures on land. They had developed thicker skin and stronger skeletons than amphibians, and they were able to take advantage of a variety of new food sources and niches.

As early as the Pennsylvanian Period, some amphibians had evolved new adaptations to allow them to live on land continually. Hard-shelled or leathery and self-contained eggs freed the reptiles that evolved from these ancestors of the need to return to the water to reproduce. Increasing aridity of the continent during the Permian gave an advantage to any creature that did not depend on abundant water for its survival. Amphibians of the Permian were large creatures, up to six feet long. They stayed mostly in the coastal, swampy regions, while the reptiles dominated the uplands.

In the deserts that formed the Coconino and De Chelly Sandstones from the Grand Canyon, Monument Valley, and Canyon de Chelly regions, tracks of invertebrates and vertebrates crisscross the rocks. More than 270 million years ago, spiders, scorpions, beetles, and millipedes left tracks in the fresh sand of the wind-blown dunes. Reptiles, perhaps crossing from one waterhole to the next, left their tracks alongside the spiders and scorpions. Most of the fossilized tracks are preserved on the ancient dune faces, which had perhaps been wetted with a light morning dew or rain to preserve the detail of the footprints. Several different types of tracks and burrows have been described; since no bones have been recovered, it is difficult to determine how many different animals these may represent. It is clear, however, that rather than being desolate regions, these ancient coastal sand deserts were places where some forms of life existed and perhaps even thrived.

Toward the end of the Permian Period, shallow seas inundated the western portion of the Colorado Plateau once more. In these seas, we find the last of the great Paleozoic marine communities. As ever, brachiopods were the most common shellfish. Clams and oysters have since taken their place. Brachiopods thrived in the warm, clear waters of these last Permian seas, surrounded by crinoids, bryozoans, mollusks, and sponges, as well as coiled cephalopods, fish, and sharks.

The plentiful fossils of Arizona and Utah's later Permian sediments on the plateau represent the last record of a community that had endured almost three hundred million years.

At the close of the Permian, more than 90 percent of all the world's species disappeared, part of a huge mass extinction known poetically as the Time of the Great Dying. Nearly all marine species and around 75 percent of all land species on the planet became extinct. After this, marine communities took on a distinctly modern look. The trilobites and most of the bony fish of that time disappeared. Although a very small number

of crinoids and brachiopods exist in a few places in today's oceans, clams, snails, sea urchins, sea stars, and sand dollars have filled their niches. These changes at the end of the Paleozoic were not limited to the seas; life on land took a new turn as well.

LATE PERMIAN

⅓ actual size

⅓ actual size

Late Paleozoic seas of the western Colorado Plateau were rich in creatures such as brachipods (above right) and sharks (right).

TRACKS THROUGH TIME

Tracks are important fossils that until recently have been overlooked in the hunt for exciting bones. Tracks preserve an animal's behavior and can lead us to conclusions about the animal's size and posture, and the speed and direction it was moving. Some tracks show where an animal has been injured in the past, and how it has recovered. Tracks may even be able to help us interpret an animal's habits or activities, such as hunting, migrating, socializing, or just hanging out at home. A track-site in the Permian Cutler Group rocks of southeastern Utah illustrates the predatory sail-finned reptile *Dimetrodon*

snapping up its smaller prey in mid-stride.

Many of the layers of rock on the Colorado Plateau contain tracks of amphibians, reptiles, worms, beetles, millipedes, spiders, scorpions, birds, and mammals. Dinosaur tracks are perhaps the most exciting to see, and certainly they are the most dramatic in appearance. We've learned much from studying them. Many archaic and entrenched ideas about dinosaur behavior and appearance have changed through the study of their tracks.

Numerous dinosaur track-sites occur throughout the Colorado Plateau, and more are being discovered every year. It is estimated that in

eastern Utah alone there are more dinosaur tracks than anywhere else in the world. A site outside of St. George, Utah, preserves hundreds of prints in the Moenave Formation. These are among the best-preserved examples of *Dilophosaurus*-like and *Coelophysis*-like tracks ever recorded and include evidence interpreted as swimming and sitting marks. More tracks exist in these ancient red muds across Navajo Nation lands. At the mid-Jurassic–aged Moab Megatracksite, on the boundary of Arches National Park in southeastern Utah, researchers estimate that there are several million three-toed carnivore tracks

within a hundred square miles. This is the only Jurassic megatracksite yet documented in the world. The millions of tracks ranging from ostrich-sized to a large *Allosaurus*-like theropod were probably the result of animals meandering back and forth over a long time across the muddy coastal plain.

The Late Cretaceous environment that would one day yield economically important seams of coal was also home to dinosaurs that left their tracks in the mud of green swamps near the seashore. Thousands of tracks have been recovered from coalmines across the plateau.

Though science has only recently acknowledged the

importance of dinosaur tracks, people have been examining them for a long time. Pictographs created by the ancient Fremont Indians more than a thousand years ago in what is now southeast Utah's Grand Staircase–Escalante National Monument, may be the earliest known example of dinosaur track documentation!

Studying tracks can prove challenging, as many of the techniques used to copy them are harmful to the fossils themselves. Rather than pouring plaster casts, or making latex rubber peels, some researchers are opting to use photogrammetry. This photographic technique allows a picture to be taken in which all distortion can be removed, thereby giving the researcher a perfect replica of the track.

Left to right, Fremont Indian pictographs of dinosaur tracks; reptile tracks in sandstone; track in Utah destroyed by plaster; carnivorous dinosaurs left their tracks over 150 million years ago in what is now southern Utah.

AGE OF REPTILES

THE MESOZOIC

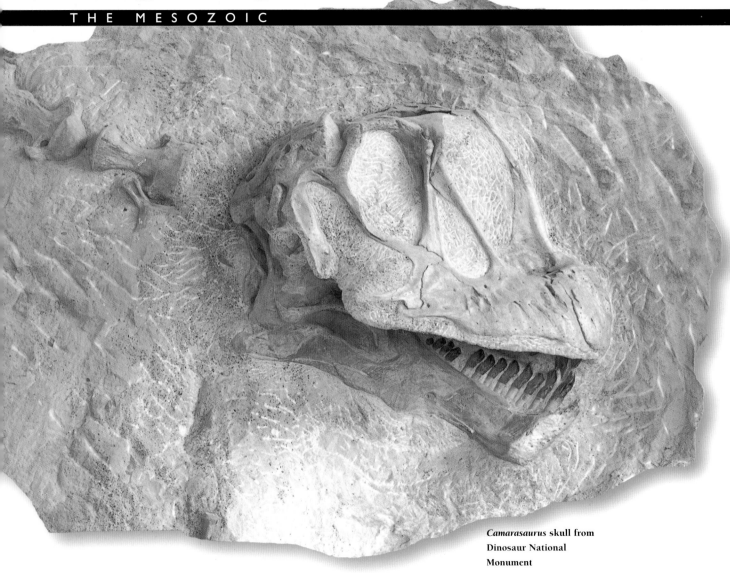

Camarasaurus skull from
Dinosaur National
Monument

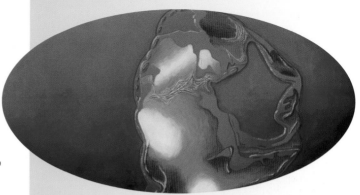

A s the crustal plates of the earth collided to form the supercontinent Pangea in the dying days of the Paleozoic, the effects were felt around the organic world. On land, the last great coal swamps of the Pennsylvanian Period were obliterated; the climate of the newly assembled supercontinent favored plants that did not need as much moisture. Reptiles, which diversified into abundant new niches, dominated vertebrate life on land. Many of these events are preserved in the Mesozoic sediments of the Colorado Plateau to make one of the world's finest records from this period.

LARGE RIVERS AND SMALL DINOSAURS—THE TRIASSIC

During the Early Triassic, as Pangea moved north, the polar ice cap melted and the worldwide climate became warmer overall with milder seasonal variation.

The plateau region sat just above sea level, slightly north of the equator. It had broken free of Pangea and was traveling northwest towards its current position. The sea had by now retreated all the way to Nevada. The deep-red Moenkopi Formation preserves the remains of large amphibians that lived in the sluggish streams that wandered across this lowland.

Giant amphibians of the Early Triassic reached lengths of five to six feet. Their flat, spade-shaped heads were sometimes a quarter of their body length, and their mouths were filled with many deadly sharp teeth.

On river floodplains walked an enigmatic creature that left only its tracks, called *Cheirotherium*, that look remarkably like human hands. The track maker seems to have been a dog-sized, carnivorous reptile that wandered the coastal plains in search of food.

The Chinle Formation is known throughout the

PANGEA

At the beginning of the Mesozoic the Colorado Plateau region was part of the great supercontinent Pangea. The plate on which the plateau rests soon broke free and began to drift northwestward toward its modern position. Volcanoes formed and mountains crumpled along the west coast as a result of plate collisions, shedding layers of mud and sand. These western highlands provided barriers to seas that periodically invaded the land, forcing the waters to come around from the north and even the east.

The Chinle Formation in northern Arizona's Painted Desert

Northern Arizona's Lithodendron Wash flows through the painted badlands of the Chinle Formation (above) where jewel-like petrified wood (right), ferns (left), and horsetails grew along the banks of ancient rivers that flowed across the Colorado Plateau during the Triassic.

TRIASSIC

Colorado Plateau, and in it researchers have discovered the remains of a remarkably diverse ecosystem. The environment of the Colorado Plateau during this time was vastly different from that of today. When one drives through the desolate and strangely lovely pastel badlands of northern Arizona's Painted Desert, it is difficult to imagine the Triassic world. Instead of a desert, we must envision a sub-tropical, seasonally humid land with a vast and braided, muddy river system flowing from highlands near the Panhandle of Texas to the coastline in central Nevada.

Abundant plants grew along the banks of these rivers and in the uplands that fed the system. Barrel-shaped palm-like cycads, ferns, and the giant horsetail rush *Neocalamites* formed the subtropical forest that grew along the river-banks themselves. The most obvious fossil plants from these shales are the giant conifers *Araucarioxylon* which reached heights of 150 feet or more. *Araucarioxylon* may have lived right along the rivers in addition to being washed in from elsewhere with floodwaters and deposited in the sediment. Petrified trunks of these trees have been found throughout these Triassic sediments on the plateau, but nowhere as abundantly as in Petrified Forest National Park.

By the Late Triassic a new member of the terrestrial ecosystem had appeared. In the tropical to subtropical environments of this time dinosaurs began their long reign over the land. The Chinle Formation holds the remains of some of North America's earliest dinosaurs. *Coelophysis* is one of the plateau's more famous carnivores. This graceful six-to-eight-foot predator walked the Chinle rivers at northern New Mexico's Ghost Ranch and at Petrified Forest National Park in eastern Arizona. It walked on two legs and had small arms used for grasping prey. This swift, agile creature with lightly built hollow bones had a long head with large eyes and sharp, serrated teeth. It moved about the Triassic uplands with its long, slim tail held out behind as a counterbalance. *Coelophysis* ate small lizards, insects, even others of its kind, as evidenced by an adult fossil from New Mexico with juvenile *Coelophysis* bones preserved in its stomach region. *Coelophysis* may have spent much of its time in packs of several individuals. A spectacular find at Ghost Ranch preserves literally thousands of

Many other animals shared the forests and waterways of the Petrified Forest with *Coelophysis* (left).

Chameleon-like lizards scurried about under the feet of *Desmatosuchus*, an omnivorous reptile. Crocodile-like phytosaurs were dangerous predators in and near the rivers. Dinosaurs would one day dominate such an ecosystem, but during the Late Triassic other animals reigned and dinosaurs stayed out of the way.

complete or nearly complete *Coelophysis* skeletons, appar-
ently the victims of a mass catastrophe.

Large amphibians joined these early dinosaurs along
Triassic waterways. *Metoposaurus* spent most of its time in
the water; perhaps occasionally lumbering out onto the dry
land of the riverbanks. While *Metoposaurus* was a predator,
eating fish and other small creatures, the most fearsome
predators in the waters of this Late Triassic kingdom were
the phytosaurs. These creatures resembled crocodiles, but
had nostrils near their eyes instead of at the tip of the

Dilophosaurus was the
earliest large meat-eating
dinosaur (far right). It stood
about six feet high at the
hip, was about fifteen to
twenty feet long, and probably walked much like
Coelophysis, with its backbone held parallel to the
ground and its tail out-
stretched for balance. While
it was carnivorous like
Coelophysis, it was much

more powerful due to its
size. Two parallel crests, that
may have been used for
display during mating rituals,
stood along its head.

Fossilized tracks of
Dilophosaurus can be seen
today near Tuba City,
Arizona.

snout. Some forms reached lengths of thirty feet. Phytosaurs could probably move swiftly and would have been supremely dangerous in water.

Formidable predators weren't the only interesting animals in this Late Triassic ecosystem. Freshwater crabs, clams, and snails lived in the waterways, and worm traces hold evidence of soft-bodied creatures not otherwise preserved as fossils. We may even be seeing the origins of bees and wasps, as well as crayfish and possibly termites in these sediments. Trace fossils interpreted as bees' nests have been found in petrified logs at Petrified Forest National Park. If these features were indeed formed by bees, it would be the earliest true evidence known of bees anywhere on the planet and the earliest evidence of social behavior in insects.

What would make this even more interesting is that these early bees would have evolved and existed without flowers, which did not appear until later in the Mesozoic.

The earliest mammals also appeared at this time, tiny shrew-sized descendants of mammal-like reptiles. The oldest mammal in the world comes from the Chinle Formation–equivalent of Texas, and the earliest known mammal trackway was found in the Chinle Formation in western Colorado. Its rodentlike track is evidence that mammals probably existed in relative abundance by this time. These early mammals

An early Mesozoic mammal

were not at all like most modern mammals, and instead had more in common with today's duck-billed platypus.

Throughout the Mesozoic, dinosaurs diversified into many niches and different forms. They became larger in body size and expanded worldwide. For over 150 million years, the dinosaurs were the ruling vertebrates on land, keeping early mammals in check as small, nocturnal forest dwellers.

Dinosaurs hardly resembled modern reptiles. Most researchers place them in their own class, apart from the reptiles. Instead of the sprawling gait of lizards, turtles, and amphibians, dinosaurs walked with their legs underneath their bodies, like mammals or birds. This allowed for more efficient and sustained movement. There is evidence that some dinosaurs may have been partially or wholly warm blooded, unlike modern reptiles, whose body temperature depends largely on the external environment. Dinosaurs

Research in the Morrison
Formation at southwest
Colorado's Dry Mesa Quarry
has helped reconstruct the
ancient environments of the
giant Jurassic dinosaurs that
wandered the Colorado
Plateau.

may have shared other characteristics with modern birds and mammals, such as herd behavior, long-term care for offspring, complex social structure, even complex communication—attributes that made dinosaurs highly efficient and contributed to their extended success. In fact, most modern analyses place birds within the class of dinosaurs.

ON THE TRAIL OF THE GIANTS—THE JURASSIC

By Jurassic times, dinosaurs on the Colorado Plateau were common and widespread. From the lake, river, and floodplain sediments of the Moenave and Kayenta Formations in northern Arizona and southern Utah, hundreds of three-toed tracks have been discovered. Some of these tracks probably belonged to a meat-eating dinosaur called *Dilophosaurus*.

Great sand deserts swept through the region in the Early Jurassic, leaving the Navajo Sandstone behind. Although the environment was not conducive to fossilization, some remains have been found in the moist interdune pond (oasis) deposits within the sandstone. These include small freshwater invertebrates known as ostracods, a tantalizing, fragmentary skeleton of a small carnivorous dinosaur called *Segisaurus*, even a crocodile-like reptile. Tracks are more common than skeletal remains in the Navajo Sandstone. Precursors to the giant long-necked dinosaurs to come walked the dunes near what would become Lake Powell in northeastern Arizona. These tracks are surrounded by those left by a tiny, three-toed carnivore, which are not much larger or different in appearance than turkey tracks.

One more incursion of the sea left the Summerville Formation above the great Jurassic desert sands. This grey and tan gypsum-rich siltstone is thought to have been left along the tidal flats of a desert-bordered sea that stretched

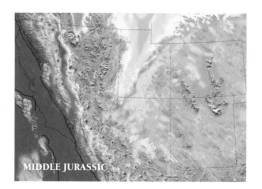

MIDDLE JURASSIC

into Canada. At the top of this formation, during the last gasp of this sea, hundreds if not thousands of tracks of flying reptiles have been found at several sites in eastern Utah, Colorado, and Wyoming in what is interpreted as a shallow embayment of the sea. The tracks of these pterosaurs tell us they were walking on all fours on the shoreline, and some prints even suggest that while in the water they paddled with their hind feet, like sea birds do today.

During the Late Jurassic, many of the dinosaurs had evolved to giant size. For more than a hundred years now, scientists have been finding fossils of these animals and many others in one of the most remarkable rock layers in the western United States: the Morrison Formation. This colorful shale and sandstone mixture erodes into painted badlands much like the Chinle Formation. Throughout the northern half of the Colorado Plateau, this formation has yielded thousands of fossils, ranging from invertebrates to tiny microvertebrates, plants of all kinds, mammals, and massive dinosaurs. The unique communities of animals portrayed by the fossil remains of the Morrison were easily as complex and varied as the great mammalian faunas of the modern African grasslands. To date, the Morrison has

The great herbivores
of the Jurassic needed
help to digest the enormous
amounts of vegetation they
consumed. Scattered throughout the formation are polished
gizzard stones, or gastroliths,
that the animals swallowed
to help grind
their food.

yielded twenty-five to thirty different dinosaur species alone. Clearly, not all of these dinosaurs lived at the same time—the Morrison spans several million years—but this is still an outstanding variety, even when averaged over time.

In a remarkable pocket of sediment on the very edge of the plateau in northeastern Utah and northwestern Colorado, a jumble of bones preserves a detailed story of Late Jurassic life in that region. The bone bed at Dinosaur National Monument lies at a steep angle in the uplifted and tilted Morrison Formation, whose slopes and cliffs of variegated purples, blues, and yellows tell of great rivers meandering across a humid and seasonally wet landscape from highlands in the west and south. Animals that drowned or died on the floodplains of these rivers were washed downstream with floods, until they came to rest as the waters receded. Several dinosaur species and hundreds of tons of bone have been excavated from the quarry here.

The immense long-necked sauropod dinosaurs, such as *Diplodocus*, *Camarasaurus*, *Apatosaurus* (formerly called *Brontosaurus*), *Barosaurus* (with the longest neck of all long-necks found in the Morrison Formation), and the rarer *Haplocanthosaurus* lived on the lowlands. Instead of browsing the treetops as their long necks would suggest, many of these dinosaurs probably munched watery streamside vegetation and may have pushed over trees like modern elephants when they wanted leaves. These giants attained unbelievable proportions: seventy-five feet long, forty feet high, and weighing thirty tons or more. In other areas of the plateau, including the Dry Mesa Dinosaur Quarry in western Colorado, giant dinosaurs have been found that rival even the blue whale in size: *Seismosaurus* and

Painted badlands of the
Morrison Formation

Utah's state dinosaur
Allosaurus

Supersaurus may have attained lengths exceeding 120 feet!

The long-necked giants used their size for protection from predators, but other dinosaurs had different means. Using the deadly spikes on its tail, *Stegosaurus* fought off the predators *Torvosaurus*, *Allosaurus*, and *Ceratosaurus*, while *Camptosaurus* and *Dryosaurus* relied on their speed.

Dinosaurs weren't the only creatures that inhabited this ecosystem, although they were certainly the most obvious. Freshwater clams and turtles abounded; snails, fish, salamanders, lizards, frogs, and crocodiles thrived in the slow waters of the rivers or hung out in ponds and lakes. Early mammals scurried about underfoot; their fossils indicate that mammals were beginning to diversify, although they were still small, mostly insectivorous or herbivorous ground- and tree-dwelling creatures.

Although the work in the Morrison has been going on since the 1870s, new fossils are constantly being found. The most complete sauropod dinosaur found anywhere in the world, a new form of *Haplocanthosaurus*, was found in the late 1990s near Dinosaur National Monument. This skeleton is 80 percent complete.

The Book Cliffs overlook southern Colorado's Grand Valley.

Allosaurus display at the Cleveland-Lloyd Dinosaur Quarry

In excess of twelve thousand bones have been excavated at the Cleveland-Lloyd Dinosaur Quarry near Price, Utah. Here, animals became trapped and sank in the mud surrounding a Jurassic spring-fed pond. Of the nearly seventy individual dinosaurs found here, more than forty were of the carnivore *Allosaurus*, Utah's state fossil and the best understood carnivorous dinosaur in the world, due in large part to the remains found at this site. This thirty-foot-long terror was the dominant predator in the Late Jurassic ecosystems of the Colorado Plateau. The find at Cleveland-Lloyd is especially important because it contains individuals of all ages, which helps us interpret dinosaur growth habits and life stages.

Near Grand Junction, Colorado, the Fruita Paleontological Area sits nestled in the Grand Valley in the shadow of the Book Cliffs. This Morrison site has proven to be one of the best in the world for finding articulated (with the bones still in original position), or semi-articulated micro-vertebrates. While the site has yielded large dinosaurs, it is a remarkable resource for smaller creatures. It holds dinosaur eggs, hatchlings, lizards, snakes, frogs, fish, small pterosaurs, and very small mammals. *Fruitachampsa*, a crocodile about three feet long, was a speedy little fellow that lived on dry land. One minuscule mammal was found curled up in a circle the size of a quarter. Some of the mammals from this site may even prove to be ancestral to many later mammal lineages.

In New Mexico, the Morrison Formation has yielded features that are being interpreted as the largest fossil termite nests known. These giant twenty-foot-high pillars, riddled with interconnected galleries and chambers identical to modern termite nests, may extend more than a hundred feet below the ground. Smaller nests have been found in Colorado and Utah as well. These insects may have played a major role in recycling organic material during the Jurassic as well as providing one source of food for some otherwise herbivorous dinosaurs.

Plants recovered from the Morrison Formation suggest warm, wet, temperate forests and plains on the uplands near the rivers. At least 225 different types of plants have thus far been recovered from research in the Morrison, and there are probably more. Conifers are common, and are preserved as colorful petrified logs in many Morrison outcrops. Massive horsetails up to ninety feet tall grew among the taller trees. Ginkgoes, giant tree ferns, and tall palm-like cycads made up the middle story of the forest, and the ground cover consisted of ferns, short cycads, small horsetails, and mosses. This was a rich primeval environment with abundant food sources to be divided among all the herbivorous giants of the day. Nothing would look very familiar to our eyes, so used to seeing the flowering plants of the modern age.

As many paleontologists on the plateau are quick to point out, the Morrison Formation is not the only layer from which remarkable fossil finds have been made. While the Morrison is extremely widespread and famous for the number and diversity of fossils it contains, other formations are proving to be extraordinarily important. Many of these new finds are

Clockwise from left, fossil seed fern, seed, and conifer cones, from the Morrison Formation

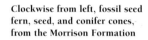

2x actual size

actual size

½ actual size

actual size

The environment of the Morrison Formation on the Colorado Plateau

coming out of Cretaceous-age sediments, especially the Early Cretaceous Cedar Mountain Formation.

ENDINGS AND NEW BEGINNINGS—THE CRETACEOUS

With the beginning of the Cretaceous Period, the world experienced a profound change. A new group of plants known as the angiosperms, or flowering plants, had arisen and would soon dominate the plant world. Because of their relationship with pollinating insects, which also diversified during this time, flowering plants had an advantage over the seed-bearing plants that relied only on wind or the chance encounter with an animal to disperse their seeds. While the cycads, ferns, tree ferns, and conifers still existed in abundance early in the Cretaceous, the first unquestionable angiosperms on the Colorado Plateau appear in the Cedar Mountain Formation. By the end of the Cretaceous, the angiosperms would comprise 90 percent of the plants in the world.

In southeastern Utah near Canyonlands and Arches National Parks, paleontologists have made some of the most important discoveries of recent years in this formation. Though not widespread geographically, the Cedar Mountain Formation, once thought to be essentially barren of fossils, has single-handedly filled in our knowledge of life on the Colorado Plateau 110 to 120 million years ago.

Many of the animals from this layer make their first appearance in North America or even globally in this region. The discovery of early marsupials, including a marsupial-like form known as *Kokopellia*, supports the idea that all marsupials may have arisen first on the Colorado

LATE CRETACEOUS

Plateau and then spread to other continents. These early forms were the size and appearance of rodents—it took a while for them to evolve into the familiar kangaroos, opossums, and koalas of today!

Several dinosaurs new to science have also been discovered in the formation, including new members of the giant brachiosaurs, titanosaurs, and camarasaurs. Several families, including the duck-billed dinosaurs and the carnivorous tyrannosaurs, appear for the first time. *Eolambia* is the oldest known duck-billed dinosaur and a form found only in Utah.

The armored dinosaurs were highly specialized creatures, resembling small tanks covered with bony armor and complete with artillery. At the end of the tail was a bony club, used to dissuade anything that tried to take a bite. *Gastonia* and other armored dinosaurs, some the size of small elephants, shared the lakes, rivers, and floodplains of the time with *Nedcolbertia*, a six-to-eight-foot-long carnivorous theropod dinosaur known only from Utah.

Some Cedar Mountain carnivores weren't small, however. The largest raptor ever found comes from the Cedar Mountain Formation: *Utahraptor*. This fearful and quite probably feathered predator was a much larger relative of the famous *Velociraptor*, and sported a twelve-to-fourteen-inch slashing claw on the inner toe of each foot, used for gutting its prey.

Most of the long-necked herbivores discovered in the Cedar Mountain Formation are fairly small when compared to the giants that preceded them. At the Dalton Wells quarry, near Arches National Park, a group of mostly juvenile titanosaurs died, were trampled by scavengers, and then transported in ash-filled mud flows to their final resting place. Nearby, paleontologists discovered a possible

nesting site. In the same formation at Dinosaur National Monument, paleontologists uncovered one of only three complete sauropod skulls known worldwide from the Cretaceous.

Plants from the Cedar Mountain Formation include conifers, cycads, the tree fern *Tempskya* and early angiosperms from a tropical family of woody vines and climbers. In overall makeup, the plants are similar to those found in the older Morrison Formation, suggesting relatively humid, sub-tropical-to-warm temperate conditions, with rivers carrying large conifers down with floods to rest in the sediments of the channels and floodplains. Crocodile and

Barrel-shaped cycads were common throughout the Mesozoic on the Colorado Plateau.

⅙ actual size

⅒ actual size

actual size

Turtles and snails are found in the Cretaceous-aged sediments of Grand Staircase–Escalante National Monument (below) and Mesa Verde National Park (below right).

turtle remains, as well as bones from a crocodile-like creature known as a champsosaur, tell us that this environment was rich and diverse.

Sediments from the Late Cretaceous are abundant on the Colorado Plateau. During this time, the climate was moderate and warm, and the seas once again invaded the land after almost one hundred million years of absence. The vast, shallow Western Interior Seaway stretched north to south across the interior third of the continent, from the Arctic Circle to the Gulf of Mexico. The edges of this sea provided swampy riverside and coastal habitat for dinosaurs, mammals, and invertebrates for millions of years. The Late Cretaceous formations of the Colorado Plateau record the advances and retreats of the coastline as the sea grew and shrank over time.

Rivers near the shoreline of this sea deposited the sandy Dakota Formation. As the Western Interior Seaway

grew, the Mancos and Tropic Shales were deposited in deeper waters offshore. Oysters (and even their pearls!), snails, clams, crabs, and coiled squidlike ammonites record the invertebrate life of the time, sometimes in a spectacular manner. In western Colorado thousands of the floating crinoids *Uintacrinus* were preserved together in a remarkable death assemblage in the Mancos Shale. Perhaps they died because of a sudden change in sea-water temperature or chemistry. Both formations are widespread across northern Arizona, southeastern Utah, and southwestern Colorado. The shale forms broad, gentle, dull-grey badlands near the Hopi Mesas in Arizona, Grand Junction and Mesa Verde, Colorado, and Capitol Reef, Utah. The Dakota Formation forms a golden-brown sandstone cliff underneath the grey shale. These formations are rich with coastal plain, shoreline, and marine fossils. The earliest undisputed bees left their burrows in the sands of the Dakota Formation and

¹⁄₂₀ actual size

Thousands of crinoids floating in the Cretaceous seas of the Colorado Plateau died together and were preserved in the Mancos Shale of Colorado.

One of the earliest recognizable flowers was found in the Dakota Formation in Nebraska.

2x actual size

Fossil fern from the Dakota Formation

actual size

dinosaur tracks are extremely common here. Examination of plants found in the Dakota Formation indicates a more modern flora, and suggests a still, warm, and subtropical-to-tropical climate.

The waters of this Cretaceous sea harbored many invertebrates, sharks, and rays as well as short-necked plesiosaurs, cousins to the long-necked, paddle-finned marine reptiles known from elsewhere in the world. The closest short-necked plesiosaurs found off the plateau are in Kansas. These reptiles sported fearsome three-and-one-half-foot jaws and may have ambushed squid, fish, turtles, and sharks from deep water.

Researchers in southern Utah excavated a highly unusual dinosaur near the plesiosaurs. Therizinosaurs are bipedal, quite possibly feathered, land-dwelling oddballs that exhibited characteristics of both carnivores and herbivores, and had long, scissor-like claws on their hands. Strangely enough, this animal's final resting place is at least 100 miles offshore in this ancient sea! Therizinosaurs of this geologic age are known mainly from China.

The Straight Cliffs, Wahweap, and Kaiparowits formations in southeastern Utah's Grand Staircase–Escalante National Monument and the Fruitland and Kirkland Formations in the San Juan Basin of northwestern New Mexico are filled with stories from the Late Cretaceous, as are the sandstones and shales exposed at Mesa Verde National Park, Black Mesa in northern Arizona, and in central Utah near Price.

The coastline, rivers, and floodplains of the Late Cretaceous Colorado Plateau grew thick with lush vegetation. After the monotonous green of the pre-Cretaceous

JACKETING A FOSSIL

In order to protect fragile bone, paleontologists often "jacket" a fossil for transport from the field to the laboratory. As the bone is uncovered during excavation, enough sediment is left in place around it to protect it from jarring and destructive tools. The sediment around the area is removed, leaving the fossil and its protective rock in place (A). Any exposed bone is protected with a thin layer of paper, over which strips of burlap or plaster-soaked canvas are wrapped, covering both fossil and rock (B). This is allowed to dry into a hard, protective cast (C). The piece is then separated from the pedestal of rock on which it stands (D), and is covered completely in plaster (E). The hard part of such an excavation is determining how to get the heavy cast-encased fossil to the truck. Sometimes paleontologists just have to put their backs into it (F).

Impression of the skin of a duck-billed dinosaur from Grand Staircase–Escalante National Monument

¼ actual size

Monument has given us the most comprehensive and continuous record of Late Cretaceous land animals from our hemisphere, and quite possibly from the world.

Discoveries from the approximately eighty-four-to-seventy-eight-million-year-old Wahweap Formation indicate the greatest known diversity of dinosaurs from this age in North America.

The duck-billed dinosaurs *Kritosaurus, Gryposaurus,* and *Parasauralophus* wandered shoreline forests of the Late Cretaceous in search of vegetation. *Parasauralophus* had a long, hollow crest that extended from its nasal passages to the top of its head. This crest may have been used to make various sounds, possibly to ward off predators or communicate with others of its own species, and may suggest a very well organized social structure. A *Parasaurolophus* find in New Mexico's San Juan Basin preserves the extremely complicated internal structure of this crest. Analysis of it has led researchers to suggest that in addition to generating sound, this structure may have been used to moisten the air in nasal passages. These dinosaurs reached lengths of more than thirty feet and seemed to be as comfortable on two legs as on all four.

More than just bones have been discovered from these sediments. Impressions of the knobby skin of a duck-billed dinosaur have been discovered in the Late Cretaceous layers of southeast Utah's Book Cliffs and in Grand Staircase–Escalante National Monument. Only a few dinosaur skin impressions are known worldwide.

Other groups make their first appearance during this time as well. The earliest member of the horned dinosaurs was discovered in western New Mexico, in rocks dating about ninety million years old. *Zuniceratops* gave rise to a

forests, plants had become colorful and diverse. Forests of oaks and willows, cypress, magnolia, palms, poplars, sycamores, grapevines, and even bananas provided new food sources and habitat for the local animals.

Trees, ferns, and other understory plants have been preserved as petrified wood and as thick beds of low-quality coal on northern Arizona's Black Mesa and Utah's Kaiparowits Plateau, as well as in seams of coal layered between sandstones and shales of the ancient shoreline. These Cretaceous everglades were home to a wealth of life, big and small, including unique dinosaurs.

Research in Grand Staircase–Escalante National

long line of horned dinosaurs, and by seventy-five to eighty million years ago, Utah and New Mexico supported a variety of these creatures in what were then swampy forests and floodplains. These herbivores lived in herds and, because of their similarities to grazing animals today, they are sometimes affectionately called "the cows of the Cretaceous." It's not hard to imagine a snorting stampede of them thundering over the land! While all horned dinosaurs sported some combination of bony neck frill and horns, New Mexico's *Pentaceratops* took this to extremes, developing highly elaborate frills and horns, perhaps as protection from predators or for use in territorial fights and mating display.

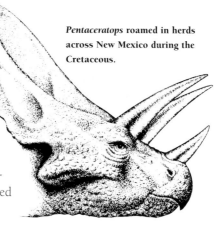

Pentaceratops roamed in herds across New Mexico during the Cretaceous.

One particularly unusual member of the late Cretaceous dinosaur community of the Colorado Plateau was a small bipedal dinosaur known as *Ornithomimus*. It may have lived much like an ostrich, running swiftly on two large hind limbs, snapping up insects, eggs, mammals, and small lizards with its long, toothless, birdlike beak.

Real birds joined birdlike dinosaurs in the forests. *Avisaurus* was a rare bird indeed, known only from one specimen in Montana and one in Grand Staircase–Escalante National Monument. This odd creature had clawed fingers at the bend in its wings. Characteristics of its feet and legs resemble a small meat-eating dinosaur. It may even have had teeth like a dinosaur, but it was definitely built to fly. It had a keeled sternum, a wishbone, and bumps on its bones that suggest attachment points for feathers.

In the Late Cretaceous forests and coastlines of New Mexico, the main predator was *Albertosaurus*, a carnivore related to the famous *Tyrannosaurus*. *Albertosaurus* stood

Fossilized hollow stump standing in life position, Fruitland Formation, San Juan Basin

Cretaceous sediments of Coal Mine Canyon at the southern end of Black Mesa (top). Similar rocks have yielded tons of coal at the mesa's northern end (bottom).

about twelve feet high and was almost thirty feet long. A huge head with long sharp teeth completed the frightening presence that must have greeted many a frantic victim. Despite its ridiculously tiny arms, its jaws and sharp-clawed rear feet no doubt made it a remarkably efficient predator. *Albertosaurus* probably gave way to *Tyrannosaurus* at the very end of the Cretaceous. In the 1990s researchers found the first *Tyrannosaurus rex* known from the Colorado Plateau in the Late Cretaceous North Horn Formation of southeastern Utah, and possibly a *Tyrannosaurus* tooth from the San Juan Basin in New Mexico.

The rest of the latest Cretaceous community that inhabited the swampy shorelines and forests of the Colorado Plateau was as extensive as its collection of dinosaurs. Mammals were still tiny, although some did reach the size of beavers. The earliest true insectivores in the region and possibly in North America appeared at this time, and marsupials had diversified into a variety of forms. Rodentlike multituberculates were the most abundant mammals during the latter half of the Mesozoic, surviving well into the Cenozoic. The tracks of pterosaurs and birds as well as the skeletal remains of turtles, rays, crocodiles, fish, sharks, amphibians, and lizards, along with an abundance of invertebrates, attest to the diverse nature of life during the latest Cretaceous.

The descendants of many, but not all, of these animals still exist today. At the end of the Cretaceous, an event occurred that was to have lasting repercussions on the history of life. There had been several extinction events throughout the Cretaceous that affected the plants and animals of the region in smaller ways. These were minor compared to the one sixty-five million years ago at the end of the Cretaceous. That extinction may have killed off as much

150 million years of ecological dominance, leaving only the birds as their descendents. The mammals survived. Theirs is the next, and latest, chapter in Earth's history.

Late Cretaceous sediments of New Mexico's Bisti Badlands

½ actual size

as 50 percent of Earth's vertebrate species in a very short time, perhaps in only a matter of years. Gone were the non-avian dinosaurs, as well as all the marine and flying reptiles and many species of invertebrates on land and in the sea, including the coiled, squidlike ammonites that had flourished in the oceans throughout the Mesozoic. In areas of North America where this has been studied, pollen and other plant fossils indicate that as much as 80 percent of all plants that existed in the Mesozoic failed to survive.

What triggered this event, or any mass extinction, is unknown. Hypotheses include global cooling caused by particulates in the atmosphere thrown up by volcanic eruptions in India or an asteroid impact in Mexico, drying of the Western Interior Seaway, or gradual climate change. Perhaps it was a combination of these factors. We do know that most of the dinosaurs became extinct after more than

actual size

Clockwise from left, fossil ammonite, ray tooth, and a carnivorous dinosaur claw from Grand Staircase–Escalante National Monument

½ actual size

MAMMALS AND

MODERNIZATION

Fossilized skull of
Hyracotherium,
also known as
Eohippus

Although true mammals had appeared on Earth around the same time as their dinosaurian competitors, they were insectivorous and herbivorous creatures ranging in size from that of a modern shrew to a beaver. As long as dinosaurs ruled the land, mammals were destined to remain on the sidelines, small and mostly nocturnal.

EARLY MAMMALS AND GREAT LAKES

With the beginning of the Cenozoic Era, the mammals' world changed. The Western Interior Seaway had drained away as the Rocky Mountains began to be uplifted, and the great reptiles that once dominated the land, sea, and air were gone. During the Paleocene (pre-dawn age) and Eocene (dawn age) mammals underwent an explosive evolution and diversification that resulted in the variety of

The colorful spires of Bryce Canyon (top) are formed of the Claron Formation, deposited in an ancient lakebed during the early Cenozoic. **The earliest horse,** *Hyracotherium* **(facing page), lived at the margins of some of these lakes.**

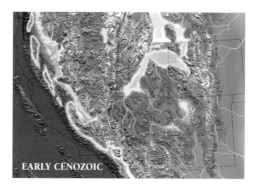

EARLY CENOZOIC

mammalian lifestyles that surrounds us today. Within a few million years of the dinosaur extinction, there were more kinds of mammals than there ever were of dinosaurs!

Cenozoic sediments from the Colorado Plateau are best preserved in the Bryce Canyon and Cedar Breaks region of south-central Utah, the Piceance Basin of northwestern Colorado, the Uinta Basin of northeastern Utah, and the San Juan Basin in northwestern New Mexico. During the early part of the Cenozoic, these basins developed large inland lakes as rivers flowing in from the south, and coming off the newly formed Rocky and Uinta Mountains ponded in the lowlands. These lakes and the streams feeding them were rich environments where early mammals flourished and proliferated. In fact, mammal fossils are so abundant that portions of these regions have produced standard reference faunas used around North America for dating rocks of that age.

The climate in the western part of the United States was warm and subtropical, and some of the more common plants that appear in these early Cenozoic formations include palm, birch, fig, magnolia, currant, ginger, heliconia, sycamore, and maple. After the initial drop in plant diversity following the Cretaceous-Tertiary extinction event, angiosperms in particular diversified with almost as much energy as the mammals.

The teeth of the earliest mammals had tiny sharp cusps for puncturing and grinding food. With small bodies and high metabolisms, they needed high-protein, high-calorie foods such as insects in order to survive. With time, mammalian teeth evolved long shearing ridges for slicing vegetation, knifelike edges for ripping meat, and tall, blunt cusps for grinding nuts, roots, seeds, and twigs.

The top predators to cross the Cretaceous-Tertiary boundary were all birds or reptiles. But despite their scarcity in the middle of the Paleocene, mammalian carnivores became quite common by late in the epoch. Grazing animals with ridged teeth designed especially for eating hard, silica-rich grass would not diversify until the Miocene, at least 40 million years into the Cenozoic, when grasslands spread.

The great lakes of the early Cenozoic, known collectively as the Green River Lakes, left a treasure trove of stories for today's paleontologists to unravel. On the northern plateau in Paleocene times, Flagstaff Lake stretched north to south from the Bryce Canyon region into the Uinta Basin. The younger Uinta and Fossil Lakes covered substantial portions of Utah, Wyoming, and Colorado in the Eocene. Today, the Piceance and Uinta Basins and the Book Cliffs of southeastern Utah and southwestern Colorado preserve sediments and fossils from these younger lakes deposited in a variety of settings: lake and lake margins, and stream and stream margin environments.

Palm frond from the Eocene Green River Formation

These lakes were long-lived features, accumulating thousands of feet of sediment over many millions of years. They record not only the evolution of life forms through time, but changes in climate as well. At the beginning of the Eocene a brief warming period resulted in more tropical conditions, after which the climate slowly cooled and became more seasonal.

A tremendously rich flora and fauna lived in and around these subtropical lakes and streams. The Eocene Green River Formation was deposited in a massive well-oxygenated lake whose algae became the abundant hydrocarbons that make the formation one of the largest oil shale deposits in the world. The layer is famous for its fossils. In Wyoming, perfectly preserved fossil fish have been found by the thousands. While fossil fish are less common in the Green River Formation of the Colorado Plateau, they do exist, along with bird feathers and dabble marks from their feeding activity, lizard skin impressions, tracks of mammals and insects, numerous plants and invertebrates, even a fossil tadpole. Bird tracks in the shallow lakeside sediment show that some had webbed feet like modern ducks and some were three-

Fossil fish preserved in lake sediments of the Green River Formation

The giant flightless, predatory bird *Diatryma*

toed like wading herons and egrets. The Eocene flamingo *Presbyornis* has been found in Colorado's Piceance Basin, and in southwest Wyoming an entire rookery of these birds was preserved in the sediment. Birds had evolved from small meat-eating dinosaurs in the Jurassic, and by the Cenozoic, they were essentially modern in appearance, with a few unusual exceptions. Eocene rocks from the Colorado Plateau yielded *Diatryma*, a giant, flightless, predatory bird from a group known as the "terror cranes." It stood six to seven feet high and sported strongly clawed feet and a powerful beak for slashing and tearing.

The formations in the Uinta Basin of northeastern Utah were deposited in rivers that fed and ultimately filled in these massive lakes. The rivers that deposited the Uinta Formation flowed from the new Rocky Mountains, while those that left the gravel, sand, and mud of the Duchesne River Formation flowed from the highlands of the Uinta Mountains. These environments hosted a stunning variety of vertebrate life. Turtles (twenty-three different kinds), salamanders, frogs, fish, and crocodiles spent their time in or very near the water's edge. Away from the water lived primitive horses, early carnivores, primates, insectivores, marsupials, rodents (ranging from vole and mouse-sized

In New Mexico, the Eocene sediments of the San Juan Basin preserve more than 180 species of mammals, most of them primitive representatives of modern groups. Most were from the ancient group that counts among its descendants all hoofed mammals, elephant-like forms, and whales and dolphins. In the Paleocene, these herbivores were the size of rabbits and cats. The primitive squirrel-sized, lemurlike primate *Cantius* inhabited the deciduous forest canopy, while the earliest horse, *Hyracotherium*, browsed on leafy vegetation. *Hyracotherium* was about the size of a fox, with four toes on its front feet and three its back feet. By the Eocene, some of these mammals reached substantial proportions. *Coryphodon* resembled a modern hippopotamus in both size and appearance.

The lemurlike primate *Smilodectes* roamed Wyoming's lush Eocene lakeside forests. A similar scene existed on the shores of Utah's ancient lakes, preserved in the Uinta Basin of northern Utah (left).

PUBLIC WORKS AND PALEONTOLOGY

In 1937, men from Franklin Roosevelt's Civilian Conservation Corps served as paleontologists at Grand Canyon National Park.

Following discovery of a bed of fossil ferns in the Hermit Shale along the South Kaibab Trail, CCC workers excavated the site and constructed a protective shelter over the fossils.

By the late 1940s the shelter was in need of repair and the fossils were being covered with debris. An interpretive shelter still exists on the site, but the fossils are difficult to find under the sediment that has covered them over the years.

TOP TO BOTTOM:

CCC enrollees excavating the fossil fern exhibit quarry on the South Kaibab Trail, circa 1935

CCC construction of fossil fern exhibit, Cedar Ridge on the South Kaibab Trail, circa 1937

The exhibit was completed in September, 1937.

creatures on up to larger than a German shepherd), even the aquatic rhino *Amynodon*, and the piglike *Achaeonodon*. More than one hundred species of mammals have been recovered from the Uinta Basin alone. Small mammals, birds, and larger ungulates left their tracks in the shoreline sands and muds.

Mammalian evolution at this time was burgeoning; the Eocene was a time of great diversity. Herbivores sported all manner of seemingly awkward headgear and reached giant proportions by the beginning of the Oligocene. None of these forms have survived into the present. They worked well for that time and those conditions, but could not survive the changing climate, and so became extinct.

As the Cenozoic advanced, the climate slowly cooled, and the mammals responded with new adaptations. Most Oligocene- through Pliocene-aged rocks have eroded from the plateau; the few that remain have yielded essentially modern creatures. Fish from the Miocene-Pliocene Bidahochi Formation of northeastern Arizona and northwestern New Mexico, although different species, are almost indistinguishable from forms found in similar lakes and streams today. We do know that during this time most modern mammalian forms evolved. The spread of grasslands in the Miocene brought about the diversification of grazing animals. In Africa, ancestors of humans were moving out of the forests and onto the savannas, where they began to walk upright, leaving their hands free for gathering food, creating and using tools, making art, and ultimately developing civilization.

The last several million years of the plateau's history have been characterized more by erosion than by deposition. During the Pliocene and Pleistocene, water and wind sculpted the landscapes that awe traveler and artist today.

While this has left us with breathtaking scenery, it has not preserved much of a fossil record. The exception to this is the Quaternary Period, the last two and one-half million years of Earth history.

GIANTS AND ICE—THE QUATERNARY

During the ice ages of the Quaternary, many now-extinct animals wandered the plateau. Mammals grew large during the Pleistocene. Some of these animals, such as peccaries, bison, antelope, bear, and big cats still exist on the plateau. Many however, disappeared about eleven thousand years ago, victims of another puzzling extinction.

Most ice-age remains on the plateau have been recovered from caves, overhangs, and pack rat nests, or middens. Limestone caves within Grand Canyon and the massive sandstone caves on the Navajo Reservation and Glen Canyon National Recreation Area have yielded bones, dung, teeth, wood, vegetation, and pollen to help reconstruct the climate during that time. In general, winters were slightly colder with higher precipitation than now, and summers were much cooler, with very little rainfall. Portions of the high mountains on the plateau were covered by small glaciers, but the great northern ice sheets of Canada did not extend south onto the plateau. Most plants existed at lower elevations than they do today. For example, conifers grew inside the Grand Canyon, and pinyon-juniper woodlands dominated what are now the low deserts.

Mammoths and mastodons inhabited the grassy lowlands and forested uplands. Mammoths were true members of the elephant family and grazed on grass, while the browsing mastodons tended to eat twigs and leaves at higher elevations. More than eleven thousand years ago, an old, diseased male Columbian mammoth died at the edge of the Colorado Plateau. When paleontologists dug the bones of the "Huntington Mammoth" out of their resting place near the Huntington Reservoir in south central Utah, they were still flexible with collagen, which is rare. But the really odd thing about this find is that it was made at 9,200 feet elevation, unusually high for a mammoth.

Another giant—the giant ground sloth *Glossotherium*—stood twelve feet high on its hind legs and spent its time with the elephant-like mastodons browsing on leaves and twigs. *Nothrotheriops*, the Shasta ground sloth, was a smaller form, standing only six feet tall and weighing in at 350 to 400 pounds.

Pack rats collect vegetation from within a few hundred feet of their homes to build nests on which they then urinate to solidify the material. These nests can last thousands of years, and give us an excellent view of changes in the vegetation of an area through time.

Camels and horses roamed the grasslands along with mammoths. Both of these animals began their evolution in the New World but became locally extinct at the end of the Pleistocene. It is only in historic times that humans have reintroduced horses to the landscape of the Americas. Other animals of the ice age, now extinct in the region, were

The earliest evidence of humans on the Colorado Plateau is about eleven thousand years old: scattered chips of stone and the occasional spear point sometimes associated with the bones of ancient animals (above).

Clovis point (left)

tapirs, shrub and musk oxen, and the giant-horned *Bison latifrons*. This larger cousin to today's bison had an impressive horn span of up to eighty inches, tip to tip. Harrington's mountain goats maneuvered nimbly around the rocks and sheer canyon walls, while giant condors and predatory birds called teratorns circled overhead.

Why are these animals no longer with us? It is true that the climate began warming and rainfall patterns shifted about eleven thousand years ago as ice sheets to the north melted and retreated. This could have affected the large animals adversely, so accustomed were they to a different weather pattern and to the food sources such a climate offered.

Another intriguing idea suggests that humans, newly arrived across the Bering Land Bridge and down past the northern ice sheets, exterminated them with hunting technology that had evolved for millions of years in the Old World. The real answer may be a combination of these two factors, or it may as yet be unknown.

The mammoth and mastodon hunters of the late ice age gave way to gatherers and hunters of small game as the climate warmed and became drier. By two thousand years ago, people had settled in villages and begun to grow corn and other crops. Humans are but the latest members of the biological community to arrive on the Colorado Plateau, the most recent chapter in a story that has lasted more than a billion years.

Ground sloths were a common member of ice-age communities of the Colorado Plateau.

During the 1930s western Grand Canyon's Rampart Cave was the site of an excavation of Quaternary fossils, including fossil sloth dung. Limestone caves in the Grand Canyon have yielded many important ice-age fossils.

AND NOW

WHERE?

From the study of life on our planet, we learn that there often is no clearly defined reason that certain creatures do not survive the passing of the eons. Life is a series of adaptations. If these adaptations prove advantageous for a particular set of conditions, a species flourishes. If conditions change and that species cannot adjust genetically it will die. Conditions can change too quickly for species to adapt. Changing climate, closing seaways, moving crustal plates, new competitors, disease, or loss of food and habitat can affect a species' ability to survive. An interesting question to ponder is the role we humans play in the story.

Are human-caused extinctions natural or unnatural? Certainly we are different from any other species that has yet lived on this planet. We have inhabited almost every habitat on land where life is possible; we attempt to make ecosystems adapt to our needs, instead of the other way around. This has led us into frontiers that other animals cannot explore—and mistakes they cannot make. But we are, after all, animals. Our culture can perhaps be seen as our adaptation to the world around us, similar in some way to those of other animals. Perhaps extinctions caused by humans are just as "natural" as extinctions caused by climatic change.

But does this justify them in any case? Can we afford indifference to the fate of our fellow creatures on this planet? Most conservation biologists believe that the current rate of human-caused extinctions qualifies as a mass extinction, which may become the largest one yet in Earth's history. It is certainly worth having an understanding of the long and complex history by which we and our fellow species came to be here. How long we remain on this planet may not be ours to decide, but we do have some control over the conditions we create for ourselves and other creatures while we are here. An appreciation of the beauty and diversity of life and the history of our place here may well help us learn how to live harmoniously with other species and take better care of our home in the process. We are inextricably entwined in this web of life and to attempt to remove ourselves from the consequences of our actions does not benefit our species or any other. This tale of life and death, evolution and extinction, change and growth on our planet is a remarkable one, and will continue long after we have become extinct or evolved into a new species.

We must remember that change has occurred and will persist despite any efforts we might make to keep things static. Managers of national parks and other public lands, indeed all of us, need to be prepared to evaluate these adjustments and their context and be flexible enough to allow natural processes on our planet and in its ecosystems to continue. And we must learn to enjoy, appreciate, and respect the intricacies and mysteries of our unique and beautiful planet and its precious cargo.

FOSSILS OF THE COLORADO PLATEAU

THE FOSSIL HUNTERS: APPENDIX A

Finding evidence of life from ancient worlds is an exciting experience. But while it may seem a glamorous "Indiana Jones" adventure, more often it is hard, dirty work. Occasionally, great finds are made by accident, but usually they are the result of much planning and careful research.

Since fossils aren't littered over the ground, paleontologists spend a good deal of time researching the rocks. Knowing the ages and ancient environments of sedimentary rocks helps in determining what types of fossils to expect in a certain area. Limestones would yield evidence of ancient seas; sandstones may be of terrestrial origin. If brachiopods are the hoped-for result, then Paleozoic limestones would be the place to look, while mammal researchers concentrate their search in Mesozoic and Cenozoic sandstones and shales. Tiny mammal bones and teeth are often collected by harvester ants to armor their anthills, and paleontologists often find important fossils by looking there.

Armed with maps, global positioning system, notebook, camera, and survey and excavation gear the paleontologist begins the search. Prospecting is a big portion of any paleontological work, and it is often hot, dry and tedious. I remember summers in the badlands of Montana. The temperature would routinely reach 115 to 120 degrees Fahrenheit. The ranchers thought we were crazy, out in the midday heat looking for "rocks." We usually ended up spending the hottest hours under the truck for shade.

Recently, paleontologists have begun pioneering work using radiological survey instruments to "x-ray" fossils in the ground. The iodide sinolation counter has been used by paleontologists working in the Morrison and Cedar Mountain Formations in southeastern Utah. While this is not a foolproof method, it has proven useful in finding even very small, slightly radioactive bones encased in less radioactive sediment.

Once fossils are found, the site is carefully described, located and marked on the maps. Then excavation can proceed. If the fossils are small and resilient invertebrates, they can be removed as is. If the paleontologists are dealing with delicate bone, the fossils are protected with plaster so that they can be transported safely back to the lab.

Once at the lab, the technician cleans and reconstructs the fossils, a job that requires substantial patience and solid knowledge of the anatomy of ancient creatures. Techniques used in the laboratory include etching with acid to remove fossils from limestone, or picking away the sediment with dental tools under a microscope. Small pneumatic vibrating needles are useful for removing very hard sediment. Tiny fossils can be recovered by washing sediment through fine screens—the silt and clay wash through but small mammal and fish teeth remain behind. Special glues and stabilizers are used during preparation to protect the fossils, and if they are particularly important specimens, latex molds and casts may be made for further study. Clues to the ancient organism's lifestyle are gleaned from the fossil itself, as well as the surrounding sediment.

One of the most exciting things about paleontology is that you don't have to be a professional to find fossils and make significant contributions to the science. Several new dinosaurs and dinosaur sites discovered in Utah recently were found by members of the general public, who have since gone on to take part in the excavations. In New Mexico, a fireman out for a hike found what has proven to be one of the best fossil quarries known from the Chinle Formation—they've pulled more than sixty-five plaster jackets out of the bone bed and they are still counting! It does, however, take training and patience to successfully remove the fossils from the field to a location for proper study. If you are lucky enough to find a fossil while you're out hiking, don't dig it out. Even if it is legal, when you do it yourself you may destroy the fossil and the scientific context, which is as important as the fossil itself. Locate it on a map (if you have a GPS, get the exact coordinates), take pictures, and contact your local museum or college geology department and local land managers. You will need to show the paleontologists where you found the fossil. You may be able to assist in its removal, and you will certainly get credit for the find.

Fossils found on private land are the property of the landowner. Fossils found on state, federal, or tribal land must be dealt with according to the rules of that governing agency. No collecting whatsoever is allowed on public land without a proper scientific permit. A comprehensive law was passed in 2003 that does for fossils what other legislation has done for archeological and historical resources on federal land. The Paleontological Resources Preservation Act requires that vertebrate and other scientifically significant fossils be collected only by qualified amateurs and professionals who obtain a permit and agree to deposit the fossils in public institutions. Additionally each state has its own laws that deal with collection of fossil material. In general, vertebrate material is protected more than invertebrate material; vertebrates are rarer, and they tend to be more scientifically valuable to natural history museums and private collectors. Even if you find a fossil on your own land, let a professional know. The knowledge gained will last as long as the science of paleontology does, and may help answer some important questions about our planet's past and future.

These days, stealing from public land is epidemic. Many sites outside parks are vandalized and illegally excavated every year. There is a sizeable industry engaged in selling fossils to collectors and museums for high prices. In many cases these fossils are obtained illegally and contextual information is lacking, making them useless for scientific study. Please don't be responsible for destroying our legacy.

If you find a fossil, if you want more information about how and where to look, or if you simply want more information about life's long history, contact your local natural history museum, college, or university geology department. The sources listed on the following pages are good places to start on and around the Colorado Plateau. The most important thing is to have fun and enjoy the adventure!

FOR MORE INFORMATION: APPENDIX B

Society of Vertebrate Paleontology
   www.vertpaleo.org

Brigham Young University Earth
Science Museum
   1683 North Canyon Road
   Provo, Utah 84602
   (801) 378-3680
   http://cpms.byu.edu/ESM

Carl Hayden Visitor Center/Glen
Canyon Dam
   Highway 89
   Page, AZ 86040
   (928) 608-6068
   www.nps.gov/glca

Cleveland-Lloyd Dinosaur Quarry
   30 miles south of Price, Utah on
   Highway 10
   www.blm.gov/utah/price/quarry.htm

College of Eastern Utah Prehistoric
Museum
   155 E. Main
   Price, UT 84501
   (435) 613-5060
   (800) 817-9949
   http://museum.ceu.edu

Denver Museum of Nature and Science
   2001 Colorado Blvd.
   Denver, CO 80205
   (303) 322-7009
   (800) 925-2250
   www.dmns.org

Dinosaur Diamond
   There are so many Mesozoic fossil
sites in eastern Utah and western
Colorado that scientists, land managers and local authorities joined to
create what is called the Dinosaur
Diamond. This approximately 550-
mile long route through the region
allows travelers the opportunity to
explore numerous museums, fossil
sites and paleontological excavations
still in progress. Contact them at:
www.dinosaurdiamond.org

Dinosaur Journey
   550 Jurassic Court
   Fruita, CO 81521
   (970) 858-7282
   www.dinosaurjourney.org

Dinosaur Museum
   754 South 200 West
   Blanding, UT 84511
   (435) 678-3454
   www.dinosaur-museum.org

Eccles Dinosaur Park
   1544 East Park Blvd.
   Ogden, UT 84401
   (801) 393- 3466
   www.dinosaurpark.org

Ghost Ranch Fossil Quarry
Ruth Hall Museum of Paleontology
   Ghost Ranch Conference Center
   US Highway 84
   Abiquiu, NM 87510
   (505) 685-4333
   www.ghostranch.org/museums/
   rhmp.php

Mesalands Dinosaur Museum
   222 E. Laughlin St.
   Tucumcari, NM 88401
   (505) 461-3466
   www.mesalands.edu/museum/
   museum.htm

Morrison Natural History Museum
   501 Colorado Highway 8
   Morrison, CO 80465
   (303) 697-1873
   http://town.morrison.co.us/mnhm

Museum of Ancient Life
   2929 N. Thanksgiving Way
   Lehi, Utah 84043
   (888) 672-6040
   (801) 768-2300
   www.thanksgivingpoint.com/
   museum

Museum of Northern Arizona
   3101 N. Ft. Valley Rd.
   Flagstaff, AZ 86001
   (928) 774-5213
   www.musnaz.org

Museum of the San Rafael
96 North 100 East
Castle Dale, UT 84513
(435) 381-5252
www.museumsanrafael.org

Museum of Western Colorado
462 Ute Avenue
Grand Junction, CO 81501
(970) 242-0971
www.wcmuseum.org

New Mexico Museum of Natural
History and Science
1801 Mountain Rd. NW
Albuquerque, NM 87104
(505) 841-2800
www.nmnaturalhistory.org

St. George Dinosaur Discovery Site at
Johnson Farm
2180 East Riverside Drive
St. George, UT 84790
(435) 574-3466
www.dinotrax.com

Utah Geological Survey
1594 W. North Temple
Salt Lake City, UT 84114
(801) 537-3300
http://geology.utah.gov

Utah Museum of Natural History
University of Utah
1390 E. President's Circle
Salt Lake City, UT 84112
(801) 581-6927
www.umnh.utah.edu

**State and National Parks and
Monuments**

Dinosaur National Monument
Dinosaur, CO
(970) 374-3000
www.nps.gov/dino

Escalante Petrified Forest State Park
Escalante, UT
(435) 826-4466
www.stateparks.utah.gov

Grand Canyon National Park
Grand Canyon, AZ
(928) 638-7888
www.nps.gov/grca

Grand Staircase-Escalante National
Monument
www.ut.blm.gov/monument

Big Water Visitor Center
Highway 89
Big Water, UT
(435) 675-3202

Escalante Visitor Center
Highway 12
Escalante, UT
(435) 826-5499

Kanab Visitor Center
Highway 89
Kanab, UT
(435) 644-4680

Petrified Forest National Park
Petrified Forest, AZ
(928) 524-6228
www.nps.gov/pefo

Red Fleet State Park
Vernal, UT
(435) 789-4432
www.stateparks.utah.gov

Utah Field House of Natural History
State Park
496 E. Main Street
Vernal, UT 84078
(435) 789-3799
www.stateparks.utah.gov

Zion National Park
Springdale, UT
(435) 772-3256
www.nps.gov/zion

READING LIST

There are more books on fossils, dinosaurs, and the Colorado Plateau than any one publication could ever list. The following is a brief selection of publications for those who would like to learn more about any of these subjects:

Chesher, Greer. *Heart of the Desert Wild: Grand Staircase–Escalante National Monument.* 2000, Bryce Canyon Natural History Association, Bryce Canyon, Utah.

Czerkas, Steven and Sylvia. *Dinosaurs: A Global View.* 1991, Mallard Press, New York.

Erickson, Jon. *An Introduction to Fossils and Minerals.* The Living Earth Series, Facts on File, Inc. 2001, Checkmark Books, New York.

Johnson, Kirk and Richard Stuckey. *Prehistoric Journey: A History of Life on Earth.* Denver Museum of Natural History. 1995, Roberts Rinehart Publishers, Boulder, Colorado.

Long, Robert and Rose Houk. *Dawn of the Dinosaurs: The Triassic in Petrified Forest.* 1988, Petrified Forest Museum Association, Petrified Forest, Arizona

Norman, Dr. David. *Illustrated Encyclopedia of Dinosaurs.* 1985, Crescent Books, New York.

Chronic, Halka. *Roadside Geology of Arizona.* 1986, Mountain Press Publishing, Missoula, Montana.

Chronic, Halka. *Roadside Geology of Colorado.* 2002, Mountain Press Publishing, Missoula, Montana.

Chronic, Halka. *Roadside Geology of New Mexico.* 1987, Mountain Press Publishing, Missoula, Montana.

Chronic, Halka. *Roadside Geology of Utah.* 1990, Mountain Press Publishing, Missoula, Montana.

Stevenson, Jay, PhD and George McGhee, PhD. *The Complete Idiot's Guide to Dinosaurs.* 1998, Alpha Books, New York.

Walker, Cyril and David Ward. *Fossils.* 1992, Eyewitness Handbooks, Dorling Kindersley, Inc., New York.

ABOUT THE AUTHOR

Christa Sadler is a geologist, guide, educator, and self-described "earth science storyteller." She has worked on rivers throughout the Southwest, Alaska, and Ecuador, and has been a guide on the Colorado River in Grand Canyon since 1988. She has worked as a naturalist and educator in Mexico, Alaska, and on the Colorado Plateau. Her research  in archeology, geology, and paleontology has included several ridiculously hot summers searching for dinosaurs in the badlands of Montana, fighting off dust storms and overly curious camels in the Gobi Desert of Mongolia, and steering clear of annoyed marine iguanas in the Galapagos Islands. She has taught introductory geology and paleontology at Northern Arizona University in Flagstaff, Arizona and works as a geology instructor for Grand Canyon Association's Grand Canyon Field Institute. In addition, she teaches geology programs for National Park Service personnel and tour guides at Grand Canyon National Park. Her business, This Earth, brings earth science programs to children around the United States, and designs earth science exercises, programs, and field trips for students in kindergarten through college. She has published stories in several anthologies, and her articles and photographs have appeared in *Plateau Magazine*, *Plateau Journal*, *Sedona Magazine*, *Sojourns*, and *Earth Magazine*. She has published an anthology of short stories and artwork by guides on the Colorado River, called *There's This River…Grand Canyon Boatman Stories*. Occasionally, she finds time to sleep. Christa lives in Flagstaff, Arizona, amid a large collection of rocks and fossils (all collected on the up and up). For more information visit www.this-earth.com.

ACKNOWLEDGMENTS

Creating this book was truly a labor of love, and my deepest thanks go to the people who recognized this, and helped it become reality: Pam Frazier, from the Grand Canyon Association, whose patience and lovely energy were inspiring throughout the project; Greer Price, formerly of Grand Canyon Association, for his support and enthusiasm; Judy Hellmich, Chief of Interpretation, Grand Canyon National Park, who supported the project from the outset; Sandra Scott, whose fine editing skills polished the book's many rough edges; and Christina Watkins and Amanda Summers for their sense of style and design.

Many fine paleontologists gave their time and energy to discuss their research. It was absolutely wonderful to see them light up when talking about their fossils. My thanks to Barry Albright, Museum of Northern Arizona; Harley Armstrong, Bureau of Land Management, Grand Junction; Sue Ann Bilbey, Utah Field House of Natural History; Ron Blakey, Northern Arizona University; Brooks Britt, Eccles Dinosaur Park; Laurie Bryant, Bureau of Land Management, Salt Lake City; Don Burge, Prehistoric Museum of the College of Eastern Utah; Dan Chure, Dinosaur National Monument; Alex Downs, Ruth Hall Museum of Paleontology; Jeff Eaton, Weber State University; Dave Elliott, Northern Arizona University; Merle Graffam, Bureau of Land Management, Grand Staircase–Escalante National Monument; Dave Gillette, Museum of Northern Arizona; Martha Hayden, Utah Geological Survey; Kirk Johnson, Denver Museum of Nature and Science; Jim Kirkland, Utah Geological Survey; David Kohls, Colorado Mountain College; Martin Lockley, University of Colorado at Denver; Spencer Lucas, New Mexico Museum of Natural History; Scott Madsen, Dinosaur National Monument; Jim Mead, Northern Arizona University; Kevin Padian, University of California, Berkeley; William Parker, Petrified Forest National Park; Peter Robinson, University of Colorado, Boulder; Eben Rose; Bryan Small, Denver Museum of Nature and Science; Lindsay Zanno, Utah Museum of Natural History and William Tidwell, Brigham Young University.

Thanks also to Brad Wallis and the reviewers, photographers, and collections managers who took valuable time to help this become an accurate and living story: Tom Pittenger, Allyson Mathis, Jim Heywood, and Carl Bowman of Grand Canyon National Park; John Ritenour, Glen Canyon National Recreation Area; Dan Chure, Dinosaur National Monument; Clint McKnight, Intermountain Natural History Association; Alan Titus, Grand Staircase–Escalante National Monument; Diane Allen, Arches National Park; Andy Karoly, Capitol Reef National Park; Michael Quinn and Colleen Hyde, Grand Canyon National Park Museum Collection; Eric Lund, Utah Museum of Natural History; Janet Gillette and Tony Marinella, Museum of Northern Arizona.

And to all my friends throughout the plateau who put me up during my travels, thank you for your love and support, and for letting me go on endlessly about fossils.

INDEX

Page numbers in *italics* refer to illustrations. An index to scientific names of organisms follows the main index.

INDEX TO SCIENTIFIC NAMES OF FOSSIL ORGANISMS

Page numbers in *italics* refer to illustrations.

A GUIDE TO COLORADO PLATEAU ROCK EXPOSURES

The nine maps at right show where rocks of different ages are exposed on the Colorado Plateau. The boundaries of these exposures are highly generalized and may not take into account smaller areas where exposed rocks of similar age may also be found. Note that colors used to indicate exposures correlate to colors used on the timeline found on the inside front cover.

1. PRECAMBRIAN

5. PENNSYLVANIAN

6. PERMIAN